Mathematical Apocrypha Redux

More Stories and Anecdotes of
Mathematicians and the Mathematical

ISBN: 0-88385-554-2
Library of Congress Catalog Card Number: 2005932231

Current Printing (last digit):
10 9 8 7 6 5 4 3 2 1

Mathematical Apocrypha Redux

More Stories and Anecdotes of Mathematicians and the Mathematical

Steven G. Krantz

Published and distributed by
The Mathematical Association of America

The Spectrum Series of the Mathematical Association of America was so named to reflect its purpose: to publish a broad range of books including biographies, accessible expositions of old or new mathematical ideas, reprints and revisions of excellent out-of-print books, popular works, and other monographs of high interest that will appeal to a broad range of readers, including students and teachers of mathematics, mathematical amateurs, and researchers.

Preface

$\mathcal{I}$t has been a pleasure to experience the warm reception that the first volume of these stories has received. The reader may understand, therefore, that I have been thereby motivated to collect more stories.

This has turned out to be easier than one might have imagined. The hard-bitten skeptic may suppose that all the best stories went into the original *Mathematical Apocrypha*. I am happy to say that that is not the case. More people have come forward with new stories, and my memory has shifted into overdrive for reviving old and forgotten stories.

For this new edition, Michael B. Henry did a marvelous job of unearthing dates and details about the subjects of various stories. His efforts add a lot to the credibility and verisimilitude of these anecdotes, and I owe him my hearty thanks.

The MAA reviewers gave my manuscript a careful read and offered many constructive suggestions and criticisms. I was frequently impressed and amazed by their knowledge of mathematical culture.

Don Albers has been an enthusiastic and proactive publisher. He did a great job of collecting photographs for this volume. Jerry Alexanderson, serving as editorial advisor for the project, was a marvel. He read many drafts of the manuscript and contributed much detail and wisdom. I am most grateful to him. Elaine Pedreira oversaw the entire editorial process for this book, and Beverly Joy Reudi supervised the typesetting and production. I am grateful to them both.

It is, as always, a pleasure to record these bits of our cultural folklore. I seek feedback and corrections from readers so that future editions may be more complete and more accurate.

SGK
St. Louis, MO

In memory of
Halsey Royden (1928–1993),
an inspiring storyteller.

And for
Paul Erdős (1913–1996),
who lived the mathematical life to its fullest.

Contents

Utter Frivolity

$\mathcal{I}$n the 1970s, Wolfgang Walter (1927–) of Karlsruhe, Germany was a frequent and enthusiastic visitor to the Math Department at UCLA. In fact Walter and his family were so taken with the place that his son ended up being a student there.

Wolfgang Walter and his host, Ray Redheffer (1921–), had many common interests in ordinary differential equations and inequalities, and he was well-liked and welcomed by all the members of the department. One day Walter and Nick Grossman (1937–) of UCLA were having a spirited conversation in the hall. What was strange about this particular chat was that Grossman was talking about John James *Audubon*, the American naturalist, while Walter was talking about the *Autobahn*, the high-speed German freeway. Each man prattled along blissfully, trenchantly unaware that the other was talking about an entirely different topic. At the end, they shook hands and each went off to teach calculus.

$\mathcal{H}$enri Poincaré (1854–1912) has been quoted as saying (at the International Congress of Mathematicians in Rome in 1908) that "Later generations will regard *Mengenlehre* (set theory) as a disease from which one has recovered." It seems that Poincaré was still reeling from the *Sturm und Drang* that had arisen over Russell's paradox, the work of Frege (1848–1925) on set theory, and related issues. Jeremy Gray (1947–) [GRA] has written a charming article considering whether Poincaré ever actually made such a vituperous statement. The final analysis seems to be that he may not have said exactly these words, but he certainly thought this thought.

$\mathcal{R}$obert E. Greenwood (1911–1993) tells me that, in 1936 or 1937, a group of Princeton mathematics and physics students (notably Ralph Boas and Frank Smithies—see [BOA]) decided to write a tongue-in-cheek article under an assumed name (H. Pétard) and to get it published in a math journal. Later, at a social function, the department secretary Agnes Fleming (1913–1986) allowed that she had received, from various sources, a number of letters addressed to one H. Pétard—asking that the letters be held for his arrival. Then, she said, she had received a letter from H. Pétard himself, saying that his arrival had been delayed and asking whether his mail could be forwarded to a new address. It seems that this group wanted a Princeton address for correspondence with the editor of the journal, but they felt that they could not in good conscience get the departmental secretary involved in their hoax.

Ultimately, the group penned the now famous paper "A contribution to the mathematical theory of big game hunting." It appears in the *American Mathematical Monthly* 45(1938), 446–447.

$\mathcal{I}$n October of 1972, Jeffrey Hamilton was lecturing at Warwick University on the subject of probability. At one point he essayed to illustrate the basic ideas by flipping a coin. The class joined him in watching the coin flip over and over and then land on the floor—on its edge! Hamilton later calculated that the chances of this happening are one in one billion.

$\mathcal{A}$ famous probabilist was teaching an elementary probability class. He illustrated the basic precepts with many of the classical examples that we all know. In particular, he taught the students that, in a room with at least 23 people, the odds are greater than one half that two people will have the same birthday. With a huge smile, he looked around the room and said, "In a room this size, the odds are tremendous that two people will have the same birthday." To his surprise, the class erupted in a huge wave of laughter. For (unbeknownst to the Professor) sitting in the front row were two identical twins.

$\mathcal{I}$n a brilliant paper called "The Complexity of Songs" (in *Comm. of the ACM* in 1984), Donald Knuth (1938–) analyzes the important problem of storing the lyrics of songs on a computer. The premise is that a song of length n, that is with n words, requires n storage spaces. But, over the centuries, various devices have been introduced to make the storage more efficient.

One of the first of these was the *refrain*. A song of length n that has a refrain may be stored in cn storage spaces, where $0 < c < 1$.

A significant improvement occurred in the Middle Ages with the song *Ehad Mi Yode'a*. This song has 13 verses $v_1,\ldots, v_{13}$. When the song is performed, the verse v_k is followed by $v_{k-1},\ldots, v_2, v_1$ before the refrain is repeated. So now a song of length n can be stored in about $\sqrt{n}$ storage spaces.

A further improvement was obtained by the farmer O. MacDonald. Since his song is well known, we shall not provide the details.

The French, always competitive, came up with their own solution to the problem. Their song *Alouette* achieves the same efficiency. The Germans contributed *Ist das Nicht ein Schnitzelbank?*.

In 1824, the true love of U. Jack gave to him a total of 12 ladies dancing, 22 lords a-leaping, 30 drummers drumming, 36 pipers piping, 40 maids a-milking, 42 swans a-swimming, and so forth. Now the complexity of a song of length n is $\sqrt[3]{n}$. When it was later pointed out that the computation is based on n *gifts* rather than n units of singing, the estimate was altered to $\sqrt{n}\log n$.

The next great improvement was American. One M. Blatz of Milwaukee, Wisconsin created a song called *99 Bottles of Beer on the Wall*. Well, we need not provide the details. Now the complexity of a song of length n is reduced to $\log n$.

Knuth concludes his paper by noting that the advent of modern drugs requires the use of still less memory to store more song. The ultimate triumph is the song *That's the Way I Like It* by Casey and the Sunshine Band. In fact the title is the only lyric, and it is repeated *ad infinitum*. Thus a song of length n, no matter how large n, can be stored in $O(1)$ space.

In a letter to the editor of *ACM Forum*, Kurt Eisemann offers further improvements. He points out that the use of "la" can result in decisive diminution of storage space. He further observes that certain native American songs consist of nothing but complete silence. Thus the estimate is …

Perhaps the details are best left to the reader.

$\mathcal{M}$ath faculty at the University of Michigan claim that they have trouble getting any consensus on evaluation of letters of recommendation for promotion and tenure cases. Seems that the problem is that several of their faculty are British. This results in dissension on the meaning of the word "quite." When a letter says, "This chappy is quite good," the British interpret this to mean that the fellow is not very good (don't believe it?—look it up in the *Oxford English Dictionary*). Repeated explanations by the American profligates are to no avail.

*P*aul Erdős (1913–1996) was once told that a friend of his had shot and killed his wife. Without blinking an eye, Erdős said, "Well, she was probably interrupting him when he was trying to prove a theorem."

*T*he movie *No Way Out* with Kevin Costner (1955–) is remarkable for a number of reasons. Costner plays a top-ranking guy in a high-security post at the Pentagon—except, in the last five minutes of the movie, we find out that he is a Russian spy. Most of us never knew that Kevin Costner could speak Russian.

A favorite part of the film takes place in the computer room at the Pentagon. The supervisor is looking over the shoulder of one of his system analysts, peering intently into the screen. At one point he blurts out, "The eigenvalues are not working. Invoke the Fourier transform!" This passage gives the film new meaning.

*I*n 1923, Béla Kerékjártó (1898–1946) wrote the book *Vorlesungen über Topologie*, published by Springer-Verlag. In the Index is a reference to a mathematician named Erich Bessel-Hagen (1898–1946). Turning to the relevant page, one finds no mention of the man Bessel-Hagen. Instead there is a picture of a torus with grotesque handles attached on the sides—something that could be interpreted as a not very flattering depiction of a man's head. In fact Bessel-Hagen was famous for having large ears that stuck out of his head.

*I*n a similar vein, Eric Temple Bell (1883–1960) published the book *Mathematics, the Queen and Servant of Science* in 1951. In the Index, there is a reference to one G. A. Miller. In fact Miller's mathematical specialty was to study and classify finite groups according to their order. So he would write a disquisition on groups of order 3, and then another on groups of order 4, and so forth. Bell did not approve of this line of inquiry, and he exhibited his dyspepsia as follows. If one turns to the cognate page in this book, one finds no explicit mention of Miller. Instead one finds a diatribe against the study of finite groups by this methodology.

Poor Bessel-Hagen was constantly an object of fun. Back in the 1930s, a number of the graduate students in Göttingen were invited to an evening party at Edmund Landau's (1877–1938). Because he was a friend of Carl Ludwig Siegel (1896–1981), Bessel-Hagen was invited too. But the hapless fellow had already purchased tickets to the movies for that evening. He did not know what to do. So he asked Siegel for advice. Carl Ludwig told him to buy some exquisite stationery and write a note to Landau explaining his predicament. Bessel-Hagen followed Siegel's advice and went to the movies. But he was never invited to Landau's house again.

Christoph Leuenberger (1968–) tells me that he recently met a guy from Cambridge, England. He told Chris that he always had a hard time because, on the one hand, he had to claim that Cambridge is much better than Oxford. On the other hand, he had to do so as if he had never heard at all of a place called Oxford …

$\mathcal{E}$minent mathematician and mathematical communicator Paul Halmos (1916–) is a man about town: he likes to dine out, and he likes to take friends to eat. Whenever he phones up for a reservation, or whenever he gives the host his name, he always says, "Fred." Asked to explain, Paul allows that it is too much trouble to explain to people how to spell "Halmos"—and they always get it wrong anyway.

Puckish complex analyst Milne Anderson (1938–) once received a paper to referee for the *Journal of the London Mathematical Society*. It was written by a hard core harmonic analyst of the old school, and concerned function spaces that he had denoted by $F_{j,k,l,m}^{p,q,r,s}$. After pondering a moment,

Zygmund and friends

Anderson wrote to the author and said he was sorry, but the journal could not consider publishing papers about spaces that have more than five indices. To everyone's surprise, the author revised the paper to comply with Anderson's admonition. And the paper now appears in the *London Journal*!

*A*nalyst Antoni Zygmund (1900–1992) used to speak of doing "centipede mathematics." "You take a centipede and pull off ninety-nine of its legs and see what it can do."

*W*alter Craig, currently a Chair Professor of geometric analysis at McMaster University, was a high school student at Berkeley High. When he was an upper classman, the school indulged in the customary tradition of electing a "Homecoming Queen"—as part of the Fall celebration of football. In today's enlightened times, most high schools elect both a Homecoming Queen and a Homecoming King. But in Walter's time there was only the Queen. His friends convinced him that the situation was unfair, and that he should run. So he did.

Walter had a wild Afro hairdo, and he would don a dashing red cape with silver trim—parading around campus to conduct his campaign. His friends distributed thousands of pamphlets in support of Walter's candidacy. *And he won!* Unfortunately the school administration took a dim view of all these shinanegans, and would not allow Walter Craig to actually *serve* as Homecoming Queen. But at least he experienced some of the glory.

*A*t a party at Eli Stein's (1931–) house, Fields Medalist Charlie Fefferman (1949–) was chatting with some people about the vagaries of life. Someone described an ad on television for a musical offering called "Great Moments in Classical Music." What the disc had was the first few bars of Beethoven's Fifth Symphony, the first few bars of *Eine Kleine Nachtmusik*, and other little popular snippets of great classics. Charlie thought a moment and said, "I can also imagine an offering called `Great Square Inches in Art'. It could have the Mona Lisa's smile in one, and the two fingers touching from Michaelangelo's painting on the ceiling of the Sistine Chapel in another, and ...."

*I*n the bad old days of the Soviet Union, the Penn State Mathematics Department was a haven for certain Russian émigrés. Today about half the department is Russian. Of course the Russian tradition of scholarship is a strong one, and Penn State has become a beacon of mathematical excellence. The Russian tradition in teaching is a rather strict and traditional one, and it does not always resonate with the students of central Pennsylvania. One of the more distinguished Russian faculty once got a teaching evaluation that began

> I knew I was in trouble when the Professor walked into class on the first day and bowed ...

*S*aunders Mac Lane (1909–2005) tells of learning Euclidean geometry in high school:

> I recall one occasion involving a theorem about a triangle: I knew that the specific shape of the triangle did not matter, and that its vertices could be lettered at will. So when I went to the board, I drew the triangle upside down, changed the letters labeling the vertices, and presented the proof at flank speed, all to the evident distress of my teacher, as some of the students (my friends) egged me on. In retrospect, it is apparent that I understood the proof from my first geome-

try class, but that I did not at all see how to communicate the proof to my fellow students; I must have been a real nuisance to the young teacher.

*J*ean Esterle (1946–), a distinguished French Banach algebraist, was once a visiting faculty member at UCLA. Now Esterle is a delightful guy, was universally liked and admired, and he spoke excellent English—but with a French accent to be sure. One of his teaching evaluations read

I don't think it's fair for the UCLA Mathematics Department to try to save money by hiring French Professors…

*R*ecently Nicolai Reshetikhin, a Russian, gave a seminar at U. C. Berkeley on the topic "Statistical mechanics, partitions and interfaces." The talk was stimulating and informative; it was well attended. At the end there was time for questions, and the distinguished applied mathematician Grigory I. Barenblatt raised his hand. He wanted to ask where this subject was going, what the long-term goals were. He phrased the query like this: "My question is rather philosophical. Imagine, if you will, that God is sitting right here …." Reshetikhin interrupted him to say, "I am atheist. I cannot imagine that."

*O*ne argument for electronic books is that students easily could carry all their texts around with them. After all, lugging a hard-copy calculus text, a hard-copy organic chemistry text, a hard-copy law book, and so forth is quite an ordeal. All these books could be put on a single CD-ROM, and the reader is hardly bigger nor heavier than a box of macaroni. What could be simpler?

But the idea has not caught on yet, and students are still lugging their books around. When I taught at UCLA, one clever student got an extra used copy of the calculus text, drilled a hole in it, and chained it to a post in the math building. That way he always had a copy of the book handy.

*W*hen British Prime Minister Tony Blair (1953–) was young he attended the Chorister School. One day his math instructor expressed puzzlement over the answer to an exam question about right triangles that Blair had

written. It was the single word "rhinoceros." "Well," said the instructor. "This is not a word that I ordinarily associate with mathematics. What ever did you mean by this?" Replied Blair, "I knew that the answer was some-thing like `hippopotamus' [hypotenuse]. But I thought that if I wrote that then you'd be upset. So `rhinoceros' was the first thing that came to mind."

*T*his yarn concerns a mathematician whose wife was not particularly math-ematical. When they went to departmental parties she was at a total loss, never knowing what to say to those around her. Finally the husband and wife devised the following scheme. They created a code. When they reached an appropriate point in the conversation, he would discreetly make a sign and she would ask "What about the infinite dimensional case?" Another sign and she would say "Does this hold in Hilbert space?" And so forth. One wonders what the signal for "It's time to go home." might have been.

*M*athematician Lee Neuwirth (1933–) was on the staff at the Institute for Defense Analyses in Princeton. He is now retired. His daughter Bebe Neuwirth was one of the stars of the popular television show *Cheers*. She played Frasier Crane's wife Lillith.

*W*hen Lewis Carroll (1832–1898) was an Oxford don he befriended Alice Liddell (1852–1934, the daughter of the Dean of Christ Church). It was for this young lady that he ended up writing *Alice in Wonderland* and *Through the Looking-Glass*.

Somehow these charming children's stories came to the attention of Queen Victoria (1819–1901), and the royal lady was highly amused. She had a letter sent to Carroll saying that Her Majesty would be delighted to read any other works that he should create. Ever the puckish one, Lewis Carroll forthwith sent her a copy of *Elementary Treatise on Determinants*.

*S*tanford Computer Scientist Donald Knuth is well known to us all for inventing the computer typesetting system T_EX, and for writing the masterwork *The Art of Computer Programming*. In point of fact, Knuth's 1963 PhD from Cal Tech, written under the direction of group theorist Marshall Hall, is in mathematics. Knuth has over 150 publications in all aspects of computer science. He is presently working on the fourth volume of *The Art of Computer Programming*. It should perhaps be better known that his *first* publication, written when he was a freshman at Case Western Reserve (and still appearing on his *Curriculum Vitae*), is "The Potrzebie system of weights

and measures," *MAD Magazine* 33 (June 1957), 36–37. In it, Knuth parodies the established system of weights and measures that we are all taught in school. For example, according to Knuth, the fundamental unit of length is the thickness of *MAD Magazine* #26 and the fundamental unit of force is the "whatmeworry." He also invents the "yllion" notation for large numbers. For instance, one myllion is one myriad myriad (10^8) and one centyllion is $10^{2^{102}}$.

For those not in the know, "potrzebie" is a word that publisher William Gaines lifted from a Polish aspirin bottle; it is the locative form of a Polish noun meaning "need."

One of the classic works of modern mathematics is Antoni Zygmund's *Trigonometric Series*. Consisting of 747 pages of dense, hard analysis, this book is the blueprint for how modern harmonic analysis on Euclidean space should be practiced. Guido Weiss (1928–) was on the scene when Zygmund was putting the finishing touches on this book. He quotes Zygmund as saying, "That's 35 papers that I could have written."

The celebrated Fourier analyst (and my mathematical grandfather) Antoni Zygmund was once on a plane when the passenger next to him endeavored to engage him in conversation. "What do you do for a living?" he queried. "I am a mathematician," replied Zygmund. "So, do you do algebra or trigonometry?" queried his neighbor in a rather snotty manner. Zygmund smiled and said, "Trig."

One day noted complex analyst Paul Koosis (1929–) stalked into the UCLA copy room, waving a document in the air. "This is it!" cried Paul. "What, what?" queried one and all. "This is it," declared Koosis, with an evil look in his eye. Finally someone demanded to know what it was. It looked like a midterm exam, pure and simple. "This is the shinola they don't know shi— from!" End of discussion.

I was recently at the annual joint meeting of the AMS/MAA in January. Wandering around the book exhibits, I ended up in conversation with the sales representative from Casio. He showed me their latest graphing calculator and lovingly described all its marvelous features. After a time he stared at my name tag and said, "So, you are at MSRI (the Mathematical Sciences Research Institute in Berkeley, where I was spending my sabbatical). How many students do you have there?" I smiled inwardly and said, "Zero". He looked chagrinned and then asked, "Well, how many faculty do you have?" I replied, "At the moment, just me." (for this was early January, and between programs at MSRI). Now he was truly abashed. He swallowed grimly and said, "I guess you wouldn't be interested in a site license."

The university in Göttingen, Germany is called "Georg-August." It was of course the home institution of Carl Friedrich Gauss (1777–1855) and Bernhard Riemann (1826–1866), and more recently of such luminaries as Hans Grauert (1930–). It is a marvelous sanctuary for mathematics. The mathematics institute at Georg-August is a separate building, and a portion of it is the "Hilbert Raum" (loosely, translated, this is "Hilbert space"). Inside this room is a statue of none other than David Hilbert. This statue is bronze, and it is hollow. It has been the tradition among students, and perhaps others, to secrete notes for assignations inside the statue.

There once was a Norbert named Wiener
Whose mind couldn't be keener
    But he'd chant and recite
    Verses so trite
That we wished he'd sing less and obscener.

—Leo Moser (1921–1970)

*W*illy Moser (1927–) is a Professor of Mathematics at McGill University. He was a great friend and collaborator of Paul Erdős. On one occasion, Moser admitted that living with Erdős was like living with a saint: he is inspiring but he is a trial. "The last time he visited," Moser says, "I set up five dinners for him with his old friends. My wife said, `He's been going to their houses for dinner for 25 years now, and he's never brought them anything. Tell him to bring some flowers or a box of chocolates.' I suggested this to him, and right away he said, `Good idea. Would you pick up five boxes of chocolates for me?' "

*O*nce Eric Milner (1928–1997) of the University of Calgary picked up Paul Erdős at the airport. He was told that he was to drive Paul to another friend's house for dinner. But there was a little problem. That "other friend" was unaware that he was hosting dinner for Paul. "But I'm sure I made arrangements to eat there tonight," said the hapless Paul Erdős. "Apparently there was some confusion," said Milner. "Well, you can come to our house instead. Don't worry, Paul, we'll feed you." But Erdős was already thinking about a math problem. Where he would eat dinner was not a matter of great concern.

*E*rdős had a friend working on harmonic analysis in Oxford, England. The poor man was hopelessly schizophrenic. When Erdős once visited him, the fellow just opened the door of his office a little bit and said, "Please come another time and to another person."

*I*n the early days of commercial airline travel, David Hilbert (1862–1943) was invited to a certain university to give a lecture. He was told to speak on any subject that suited his fancy. He announced that he would speak on "The Proof of Fermat's Last Theorem." Certainly his prospective audience was enchanted and thrilled, because no proof of this famous theorem was known to exist. It is safe to say that the event was much anticipated.

The great day arrived, and the room was packed. Hilbert delivered his lecture, but it had nothing whatever to do with Fermat's last theorem. Afterward, someone had the temerity to ask why the great man had chosen a title that had nothing at all to do with the subject matter. "Oh," said Hilbert. "That was just in case the plane went down."

A popular story among (male) Chinese mathematicians is this:

The mayor of a certain Chinese village called all the people together. The mayor stood in the South center of the village square and faced his people. He commanded that all the men who are afraid of their wives should stand in the Northeast corner of the square, and all those who are not afraid should stand in the Northwest corner. All the village men but one went to the Northeast corner, and one lone man repaired to the Northwest corner. The mayor turned to the latter fellow with a great smile and said, "I applaud you sir. How can it be that you are the only man standing in this corner?" The man stammered his reply: "My wife will not let me hang out in crowds."

*K*en Rosen is a most successful textbook author. He has penned, for instance, the best-selling undergraduate discrete math book and the best-selling undergraduate number theory book.

One of Ken's texts is used in Kuwait. He received recently a letter from the Kuwaitis protesting a feature of his book. For, in illustrating ideas of logic, he has the sentences

If $1 + 1 = 3$ then God does not exist.

If $2 + 2 = 4$ then pigs can fly.

The Kuwaitis felt that the first sentence was sacrilegious and the second affiliated the unclean pig with God.

Robert Oliver (1949–) was a graduate school classmate at Princeton and certainly one of the most impressive people in our group. His thesis led to the solution of a famous problem in group actions, and he has established himself as a powerful mathematician. He told me that, when he was a child, he had trouble telling the difference between cows and horses. His theory at the time was that the main difference was that horses had spots.

When we were in graduate school, a group of us gathered around the big table in the Commons Room at noon each day to read the *New York Times* and eat lunch. This was quite a ritual, and we stuck to it rigorously. Of course all the departmental gossip was exchanged and analyzed at that time. This is how we kept in the loop. After lunch, some of us would play Chess or Go. Eventually, we would get around to doing some mathematics.

Robert Oliver was part of the group. And, each day, he brought two tuna fish sandwiches to eat for lunch. This was his routine, day in and day out. We started teasing him about having no imagination regarding what to eat for lunch. After a time, he got tired of this abuse, and he fought back. "Well, what else is there to have for lunch?" We suggested salami, turkey, beef, ham, cheese, etc. He seemed unimpressed. But one day he came in with a pickled squid sandwich—just to show that he could be as creative as the next guy.

One day harmonic analyst Mitch Taibleson (1929–2004) sat in the coffee room. He was evidently growing more and more irritated hearing talk of departmental politics and frustrating calculus classes and repressive administrators. Finally Mitch said, "I'm moving my desk into the bathroom. I want to be around people who know what they are doing."

Max Zorn (1906–1993) was a fine old guy. He didn't get tenure at UCLA because they decided that Edwin Beckenbach was a better bet, but he spent a good career at Indiana University. And one of the most famous results in all of mathematics (Zorn's Lemma) is named after him. Even though there are so many essentially equivalent results (the Hausdorff maximality principle, the Kuratowski-Zorn lemma, and on and on), certainly Zorn's lemma itself holds a pre-eminent position in modern textbooks. It is said that, at a certain conference, Kuratowski (1896–1980) was in attendance and Zorn was the speaker. During the great man's talk, he referred to his own result, calling it in fact "Zorn's lemma". Then he paused and said, "Zorn is me, and over there sits Professor Kuratowski who proved it first."

∞

*O*ne day, when Marston Morse (1892–1977) was one of the leading lights at the Institute for Advanced Study, a visitor came to study with Morse. He gave a lecture about matters of mutual interest. Whenever there came a point when he might mention "Morse theory," he always was careful to call it something else— "the calculus of variations in the large" or some such thing. Each time, Morse would say, "What, what? What subject is that? What did you call it?" And the speaker would say, "the

Marston Morse

calculus of variations in the large." Morse went home quite frustrated that day.

∞

*I*n Spring of 1996 a conference was held in Berkeley to honor the celebrated geometer S. S. Chern (1911–2004). The event was sponsored in part by Jim Simons (1938–), a former mathematician who had opened a stock trading house (Renaissance Technologies, home of the famous Medallion Fund) and made a fortune on the market. Simons was there—he had written some famous papers with Chern in the 1960s and knew a number of the participants—and he got up at some point to say a few words. Simons is a bluff and hearty guy, and he had some fun regaling the audience with stories. He fondly recalled the days when he worked with Chern, developing the so-called "Chern-Simons invariants", and he mentioned particularly what Chern had said when he was told that Simons was quitting mathematics. "Chern said, `Well, he's no Hilbert'." Simons thought this was really touching. He said, "You know, that's really flattering. To be compared to Hilbert. He could have said, `Well, he's no Hirzebruch,' but he decided to crank it up a notch, and I've always been pleased about that." Everyone in the audience found this extremely amusing—except for Hirzebruch (1927–), who sat in the back and took it all in with a dour expression.

*W*hen I first started out in this profession (about thirty years ago at UCLA), photocopying was still considered to be something of a luxury. Most math departments still had ditto and mimeograph machines, and we produced our exams and class handouts with these rather primitive technologies. Even when I had a research paper typed (we didn't have T$_E$X in those days, and instead the department employed technical typists), I was not given photocopied proof sheets. Instead the original was equipped with an elaborate system of translucent overlays, and I was to mark my corrections with a special pen on the overlays.

Over time, the use of the photocopy machine grew and grew—at almost an exponential rate. This change had an impact on the departmental budget, and the chairman at UCLA went through some growing pains as he tried to come to terms with increased expenditures for photocopies. At one point he issued a rather pompous edict to the entire department—faculty and graduate students alike—chastising everyone for not filling out the photocopy log correctly, carefully putting name, date, and number of copies. A dyspeptic reading of his epistle was that he blamed the graduate students and other "young riffraff" for abuse of photocopying privileges.

One of the more mischievous graduate students took umbrage with the matter and wrote a satirical version of the "photocopying edict". It was found in everyone's mailbox the following Monday. It reads something like this:

TO: Faculty, TA's and graduate students

FROM: [name withheld on request], Chair

RE: Further Steps Towards Eliminating Reproduction

The problems surrounding unauthorized use of the Reproduction

Room and other sensitive areas of the Mathematics Department have become increasingly complex. It is no longer clear that the graduate students and their element are solely blameful, nor that the transgressors have confined themselves to the pirating of office supplies and the undocumented use of the Xerox facility.

Our sources indicate that an alarmingly large segment of the department have concealed criminal records, and that several others are using assumed names, and perhaps have no mathematics background whatsoever. We now believe that it is this segment of the department, in collusion with the graduate body, which are responsible for the abuse of Xeroxing privileges, and for other activities posing an internal threat to the department.

As a means of coping with the problem, I am asking that all those wishing to use the Xerox machine submit to a polygraph test. Those found deceptive will be subject to denied access to all office supplies, and/or revocation of tenure, as the situation warrants. Those of you who are innocent have nothing to fear, and will of course be anxious to comply with this new policy.

Again, I urge you all, whenever possible, to keep your personal Xeroxing to a minimum, to rely on Elaine for most of your Xeroxing needs, and most importantly, to report unauthorized or even remotely suspicious use of the Reproduction Room to me immediately, so that the troublemakers among us can be effectively uprooted.

We appreciate your understanding of the difficulties we face in keeping the graduate students out from under foot, and your cooperation in the matter is greatly appreciated.

*P*art of good scholarly writing is accuracy. And part of accuracy is proper spelling. Many of us use spell-checkers; but a spell-checker will only find an item that is *not* a word. It will not correct the writing of "weigh" when you mean "way" or "rebound" when you mean "redound." The proper names of mathematicians are a particular source of frustration, and certainly most of them do not appear in the spell-checker's lexicon. Some names, such as Tchebychev (1821–1894) and Kowalevski (1850–1891), seem to have infinitely many spellings. Ralph P. Boas (1912–1992) once wrote a poem about the matter.

Here it is:

Weep for the mathematicians
Posterity acclaims:
Although we know their theorems
We cannot spell their names.

Forget the rules you thought you knew—
Henri Lebesgue has got no Q.

Although it almost rhymes with Birkhoff,
Two H's grace the name of Kirchhoff.

The Schwarz of inequality
And lemma too, he has no T.

The "distribution" Schwartz, you see
Is French, and so he has a T.

In Turing's name—no German, he—
An umlaut we should never see.

Hermann Grassmann—please try to
Spell both his names with 2 N's, too.

If you should ever have to quote
A Harvard Peirce, be sure to note

He has the E before the I;
And so does Klein. Rules still apply

To Wiener: I precedes the E;
The same for Riemann, as you see.

But Weierstrass, the lucky guy,
Has it both ways, with EIE.

Fejér, Turán, Cesàro, Fréchet—
Let's make the accents go that way,

And as for Radon-Nikodým,
Restore the accent, that's my dream.

But there is one I leave to you,
Whatever you may choose to do:

Put letters in or leave them out,
Garnish with accents round about,

Finish the name with -eff or -off:
There *is* no way to spell Чебышёв.

…

If you would have a friend in me
Spell "Boas" thus, not with a z.

Stokes's theorem is O.K.
Stokes' theorem is another way.

Stoke's theorem has to be a joke.
Unless you thought his name was Stoke.

*I*n fact Jerry Folland (1947–), who is quite knowledgeable both about Russian and about mathematical history, has a firm opinion about the spelling of Sonja's name:

> The problem of spelling Sonya K's surname is even harder than the problem of spelling Chebyshev. I've given it some thought, and I claim that your solution is incorrect. The name is certainly of Polish origin, and Kowalewska is the correct Polish spelling with the Polish feminine ending. If only Sonya had been Polish, you'd have a winner.
>
> But she acquired the name through marriage, and she was Russian. So when she was hangin' with the homeboys, she would have spelled her name in Cyrillic with the Russian feminine ending, which transliterates (via the scheme used by the AMS) as Kovalevskaia. If you want to be a good feminist and a good Russophile, like our friend Ann Hibner Koblitz (1952–), that's probably the way to go.
>
> On the other hand, Sonya's scientific career took place in western Europe, where she wrote in languages that use the Roman alphabet and do not inflect family names for gender. She signed her first paper Kowalevsky and the subsequent ones Kowalevski. The conclusion *I* draw is that her professional name was Kowalevski, and that's how we, her professional colleagues, should refer to her.

*D*oron Zeilberger (1950–) is a very clever combinatorial theorist who was my colleague for a year (1976–1977) at UCLA while he was on sabbatical. In those days the Student Union cafeteria was the most popular place (at least among mathematicians) to have lunch. There was a great selection of sandwiches and salads, and we all looked forward to our mid-day repast.

One of the features at the cafeteria lunch was a salad bar. You could fill up a small plate for $1.50 or instead choose to fill up a large plate for $2.50.

Zeilberger caused a campuswide sensation when he wrote a letter to the student paper explaining how you could construct a cantilever system with celery sticks to turn the small plate effectively into the large plate—thereby getting a $2.50 salad for the measly sum of $1.50. The salad bar was eliminated soon after that.

André Weil (1906–1998) was imprisoned in the early years of World War II for failing to report for duty in the army. He had a trial, at which Elie Cartan (1869–1951) testified on his behalf. Weil did end up spending time in prison. This is one of those legendary events in the history of our subject, for Weil always said that this was a great atmosphere in which to do mathematics—the food was OK, and there were few interruptions. In fact he proved the Riemann Hypothesis for function fields over finite fields during his incarceration. In later years, Hermann Weyl (1885–1955) offered to use his influence to have Weil put in prison again, since his previous stay in that type of establishment had had such a positive effect on Weil's work.

André Weil

One of the most common and popular Norbert Wiener (1894–1964) stories is of a student coming to Wiener after class and saying, "I really don't understand this problem that you discussed in class. Can you explain to me how to do it?" Wiener thought a moment, and wrote the answer (and only that) on the board. "Yes," said the student, but I would really like to master the technique. Can you tell me the details?" Wiener bowed his head in thought, and again he wrote the answer on the board. In some torment, the student said, "But Professor Wiener, can't you show me how the problem is done?" To which Wiener is reputed to have replied, "But I've already shown you how to do the problem in two ways!"

Dick Swenson, who was at MIT in those days, tells this variant of the story: Wiener showed the kid the answer twice, as just indicated. Then the student said, "Oh, you mean …," and *he* wrote the answer (and only the answer) on the board. Wiener then said, "Ah, very nice. I hadn't thought of that approach."

At a recent conference in complex geometry held at a resort in Korea, we had the opportunity to sample the latest Korean snack sensation. We actually saw kids pass up pizza and McDonald's for this treat. It was steamed silkworms. Yes, folks, for about $1 you could get a cup of the little suckers (still in their cocoons) together with a large toothpick. You would spear them and eat them. Sauce was available.

Witold Hurewicz (1904–1956) had a theory of "deanology". It went like this: Let S be the set of frustrated scholars, let B be the set of frustrated businessmen, and let D be the set of deans. *Axiom: $D = S \cap B$*. And so forth. The interesting part of the theory is the method of reproduction. Sons of deans are never deans, but instead potential deans marry deans' daughters.

Hurewicz expounded this theory once at a quite formal dinner party given by a rather pompous host. The somewhat flustered hostess said, "But I'm a dean's daughter." That certainly stopped the conversation. Afterward Hurewicz looked mischievously penitent and said to John L. Kelley (1916–1999), "But what could I do? It's exactly what I meant."

F. Riesz, Rademacher, Szàsz, Knopp, Mrs. Szegõ,
Bessel-Hagen, Ostrowski, all on the streetcar tracks.

$\mathcal{G}$eorge Pólya (1887–1985) tells lovingly of his experiences as a young mathematician. In the early days Pólya was a Privatdozent in Germany. This is a financially shaky position that is something like a downgraded Assistant Professor in the U.S.A. Pólya was married at the time, and his wife developed the habit of photographing mathematicians. One day she stopped Lipót Fejér (1880–1959) in the company of three or four others, and she took a photo of them posing on the street car tracks in front of the university. She was about to take a second picture of the august scholars standing on the tracks when Fejér cried out, "What a good wife! She puts all these full professors on the tracks of the street car so that they may be run over and then her husband will get a job!"

$\mathcal{J}$ohn von Neumann (1903–1957) once owned a dog named "Inverse". René Descartes (1596–1650) owned a dog named "Monsieur Grat," which means "Mr. Scratch".

$\mathcal{L}$ike most Frenchmen of his day, Henri Poincaré (1854–1912) was in the habit of buying bread once per day from his local baker. The bread was supposed to weigh 1 kilo, but after a year of recordkeeping Poincaré determined a nice normal distribution for the weights that had mean 950 grams. He called the police and they told the baker to behave himself. One year later Poincaré reported again to the police, complaining that the baker had not reformed. The police confronted the baker and he said, "How could Poincaré have known that we always gave him the largest loaf?" Poincaré then showed the police his records for the year, which again showed a bell-shaped curve with the minimum at 950 grams but truncated on the left side.

$\mathcal{J}$ohn von Neumann (1903–1957)—whose full Hungarian name was Margittai Neumann János—was famous for his instantaneous photographic memory. Speaking of the Manhattan telephone directory he once said that he knew all the numbers in it—the only other thing he needed, to be able to dispense with the book altogether, was to know the names that the numbers belonged to.

$\mathcal{W}$hen von Neumann first came to the U.S. he spoke primarily German. But he picked up languages quickly and easily. His finesse with English was demonstrated one evening at an after-lecture party in 1934 at Harvard. Someone mentioned Lewis Carroll's *The Hunting of the Snark*. Both von Neumann and Norbert Wiener, who were standing nearby, were set afire by mention of this humorous work, and immediately began to recite, as rapidly as possible, 150 lines of the poem. They finished in a dead heat.

$\mathcal{D}$ining in Princeton has always been a trial. There are few places that offer even passable cuisine. One day a visiting French Professor asked Joe Kohn (1932–) to recommend a good place to eat dinner. Joe said, "You want to have a good dinner in Princeton? Go to the A&P and buy a loaf of bread and a baloney!"

$\mathcal{A}$dolf Hurwitz (1859–1919) was known for his complete politesse when he was in public. But, in private, he could be both witty and acerbic. He was particularly kind to his PhD students, some of whom were a bit of a trial. Even the patient Hurwitz was moved to say, "A PhD dissertation is a paper of the professor written under aggravating circumstances."

$\mathcal{J}$. R. Kline (1891–1955) liked to tell stories of other mathematicians. This one about Norbert Wiener was a favorite: One summer the Klines and the Wieners had adjacent cottages on a lake in New Hampshire. Norbert was in the habit of swimming from his dock to a small island in the middle of the lake. He would thereby convince himself that his physical prowess lagged not too far behind his mental powers. On these swims, Kline would keep Norbert company by paddling a rowboat alongside. And they would carry on a conversation while Wiener was steadfastly progressing towards his goal. Norbert always tried to keep control of the conversation, even as he was puffing and gasping towards the small land mass. On one such day, near the end of the swim, he bleated out, "Kline, who are the five greatest living mathematicians?" Quietly, Kline replied, "That is an interesting question. Let's see." He quickly ticked off four names (none of them "Wiener"). "Yes, yes, go on," spluttered Wiener. With delicate humor, Kline avoided mentioning the name of the fifth one.

$\mathcal{G}$ilbert Bliss (1876–1951), of calculus of variations fame, was one of the leading lights of the University of Chicago math department in the 1920s and 1930s. He also liked to have his moments of levity. According to Saunders Mac Lane,

> Professor Bliss liked to kid his students. One day in his lectures on the calculus of variations, he recounted his own earlier experiences in Paris. After he sat down in the large lecture amphitheatre there, an impressive and formally dressed man entered and went to the front. Bliss thought it was the professor himself, but no, it was just his assistant who cleaned the blackboards and set the lights. When the professor finally arrived, all the students stood up. At this point in his story, Bliss observed that American students do not pay proper respect to their professors. So the class agreed on suitable steps; I

[Mac Lane] was the only member owning a tuxedo. The next day, arrayed in that tuxedo, I knocked on the door for Professor Bliss to report that his class awaited him. When he came in they all rose in his honor.

G. H. Hardy (1877–1947) had very personal and definite views of almost everything—mathematical and not. He liked cats, but abhorred dogs. He loved cricket, but hated rowing. Hardy was a Cambridge man, but spent some time as a Professor in Oxford. One day someone, quite innocent of Hardy's eccentricities and strong opinions, asked him, "For which university are you in sports?" Hardy replied, "It depends. In cricket I am for Cambridge, in rowing I am for Oxford."

*T*opologist Bob McDowell (1927–) of Washington University has a theory that cancer causes smoking.

He also observes that, when you get old, everything in the world seems to be going to hell in a handbasket. But, he notes, God designed it to be that way. As a consequence, you really *want* to die.

*R*alph P. Boas was a great mathematician, and also a great citizen of the mathematical community. But he was always quite modest. He tells of giving an honorary lecture and having someone come up afterward and saying, "You make mathematics seem like so much fun." Boas's reply was, "If it isn't fun, why do it?" He allowed privately that this is too quixotic a principle to be tenable today. He also recalls a Harvard colloquium at which the speaker commented that he had not been very interested in his topic, "but sometimes one has to do some research." Boas reports that a collective shudder went through the audience.

*B*oas also tells of the rather parochial nature of AMS meetings in the early days. He recalls one particular meeting that he attended as a young man. He overheard the staff wondering what to do about Norbert Wiener, who hadn't made a room reservation. They decided that they had better hold a room for him; sure enough, presently Wiener ambled in.

$\mathcal{N}$iels Bohr (1885–1962) was a rather mathematical theoretical physicist. He had a horseshoe nailed outside his house, much as Texas ranchers do. One day a scientist friend of Bohr's asked incredulously whether Bohr really believed that a horseshoe brought luck. Bohr replied, "I don't. But I understand that it brings you luck whether you believe it or not."

$\mathcal{I}$ ran into the fiendishly witty mathematician Underwood Dudley (1937–) at a meeting a few months ago. He declared, "I have two pearls of wisdom about teaching. Are you ready to hear them?" I allowed that I was. So Woody said,

First Pearl: Nothing works.
Second Pearl: Everything works.

$\mathcal{M}$athematician Jim Carlson (1946–) of the University of Utah—and now the new Director of the Clay Mathematics Institute—likes to say that the most irresponsible thing that an adult American can do is to stand up in a crowded room and shout "Proof!" Perhaps Jim has taught calculus one too many times.

$\mathcal{I}$n 1940, over lunch, Norbert Wiener and Aurel Wintner (1903–1958), with assistance from R. P. Boas, cooked up the idea for a journal named *Trivia Mathematica*. A prospectus of their ideas appears as follows (with thanks to the reference [DUR, vol. 1, page 95]):

Announcement of the Revival
of a Distinguished Journal

TRIVIA MATHEMATICA

Founded by Norbert Wiener and Aurel Wintner in 1939

"Everything is trivial once you know the proof."
—D. V. Widder (1898–1990)

The first issue of *Trivia Mathematica* (Old Series) was never published. *Trivia Mathematica* (New Series) will be issued continuously in unbounded parts. Contributions may be written in Basic English, English BASIC, Poldavian, Peanese, and/or Ish, and should be directed to the Editors at the Department of Metamathematics, University of The Bad Lands. Contributions will be neither acknowledged, returned, nor published.

The first issue will be dedicated to N. Bourbaki, John Rainwater, Adam Riese, O. P. Lossers, A. C. Zitronenbaum, Anon, and to the memory of T. Radó, who was not amused. It is expected to include the following papers.

On the well-ordering of finite sets.

A Jordan curve passing through no point of any plane.

Fermat's last theorem. I: The case of even primes.

Fermat's last theorem. II: A proof assuming no responsibility.

On the topology *im Kleinen* of the null circle.

On prime round numbers.

The asymptotic behavior of the coefficients of a polynomial.

The product of large consecutive integers is never a prime.

Certain invariant characterizations of the empty set.

The random walk on one-sided streets.

The statistical independence of the zeros of the exponential function.

Fixed points in theorem space.

On the tritangent planes of the ternary antiseptic.

On the asymptotic distribution of gaps in the proofs of theorems in harmonic analysis.

Proof that every inequation has an unroot.

Sur un continu d'hypothèses qui est équivalent à l'hypothèse du continu.

On unprintable propositions.

A momentous problem for monotonous functions.

On the kernels of mathematical nuts.

The impossibility of the proof of the impossibility of a proof.

A sweeping-out process for inexhaustible mathematicians.

On transformations without sense.

The normal distribution of abnormal mathematicians.

The method of steepest descents on weakly bounding bicycles.

Elephantine analysis and Giraffical representation.

The twice-Born approximation.

Pseudoproblems for pseudodifferential operators.

The Editors are pleased to announce that because of a timely subvention from the National Silence Foundation, the first issue will not appear.

$\mathcal{C}$. K. Cheung (1960–) is a mathematician at Boston College. He was very happy when he bought his new house, for that made him a homeowner and an established member of the community of Boston. But he wasn't necessarily ready for all the little responsibilities of stewardship. In particular, when Halloween rolled around he was unprepared for the hordes of small children who fell upon his residence. He had no candy to give them. Not wanting to disappoint the little tykes, Cheung found a bottle of chewable vitamins on his shelf and gave each kid a vitamin. The children were suitably puzzled by this gift. One consequence was that, in subsequent years, they never returned.

$\mathcal{W}$hen I was a young mathematician, an Assistant Professor at UCLA, I would sometimes shop at the Bullock's Department Store in Westwood. This is a pretty fancy store, and I would only go there on a special occasion (such as to buy a present for my wife). One evening I was there late, and got back to my car in the underground parking lot very near to closing time. A gentleman in a nearby vehicle could not get his car started, and I offered to give him a jump-start for his battery. He was most grateful, and I helped him with courtesy and dispatch. Bear in mind that I was 23 years old at the time, and I looked like I was about 16. The grateful fellow reached for his wallet as he said, "I suppose you are a student and could use some extra cash …" Silly me. I said, "No, I'm a professor at the university." He quickly put his wallet away and left.

$\mathcal{J}$n the late 1970s at UCLA there was a graduate student who was struggling mightily with his quals. He had one last chance to pass analysis, and the magic day came. At the end of the exam, he knew it was all over. He simply hadn't made the grade. Somewhat despondent, he took himself out to a nice dinner in Westwood where he sat out on a balcony and pondered his future. He ordered a good steak, medium rare, and a glass of wine. When the waiter brought it, he sat back and pondered what a nice repast he was about to have. At that moment a bird flew overhead and relieved himself on the steak. The student took this as a sign.

*R*oger Penrose (1931–) was giving a talk at Rice University about "Minkowski space with light rays." He asserted that this was an 8-dimensional space. Eugene Wigner (1902–1995) was in the audience; he raised his hand and said that he thought Minkowski space was 4-dimensional. Penrose said, "No, this is Minkowski space with light rays. It is 8-dimensional." But Wigner was not satisfied. After a while, he queried Penrose again. And Penrose gave a similar answer. Finally Wigner threw up his hands in frustration and demanded satisfaction: "What is this secret you are keeping from us? Minkowski space is 4-dimensional!!"

*S*tanley Sawyer (1940–) was a student of Adriano Garsia (1928–) at Cal Tech in the mid-1960s. The first time he met with Garsia, the great man was busy chatting away on the phone. He would periodically grab a Kleenex from the box, stuff it in his mouth, chew it up, and swallow it. Sawyer found this behavior quite bizarre, but assumed that Garsia knew something that he didn't know. He later found out that Garsia had recently given up smoking, and that eating Kleenex was his way of satisfying the oral craving.

Garsia was always a very energetic individual. The story is that, when he was young, he would begin the day by taking some sleeping pills so that he would be more like everyone else.

*G*arsia trained Stan Sawyer to be a harmonic analyst. Indeed, he wrote a splendid thesis on limits of sequences of operators. But at some point he became bored with the world of harmonic analysis. Sawyer was on the faculty of Purdue University at the time, and one day he wandered into biostatistician Dan Hartl's office and asked him what were the two biggest unsolved problems in his field. Hartl told him, and two weeks later Stan had solved the problems. The rest is legend. Sawyer and Hartl developed an important and flourishing collaboration, and after a time they both moved to Washington University. Hartl has since gone on to Harvard, but their joint work continues.

Q: What do you get when you cross a mosquito with a mountain climber?
A: Silly! You can't cross a vector and a scaler.

Utter Helplessness

*B*oris Weisfeiler (1941–1985?) was my colleague at Penn State University. Boris was the sort of guy who always went through life with a raincloud over his head. The best advice he ever gave me (which I particularly appreciate now that I have been a Department Chairman) is, "Don't give people too many choices."

Boris found solace in taking long, solitary hikes in remote parts of the world. Among other places, he visited Alaska, Nepal, China, and Peru. He would plan these outings painstakingly—as they were usually in virtually uncharted venues. And he always managed to get into trouble. On his trek in Alaska he lost his compass and ran out of food. He wandered aimlessly for two weeks on the tundra before finally finding an outpost where he could get help. In his adventure in China he ran into similar difficulties.

Boris's undoing, however, was his excursion into Chile. All that is known about that adventure is this:

- Boris arrived in Santiago on December 25, 1984.
- He never returned. He is thought to have "disappeared" on January 5, 1985.
- The official story, endorsed by the Chilean government, is that Boris drowned at the confluence of the Nuble and Los Sauces rivers near the border between Argentina and Chile.
- His backpack and visa were found near a river.
- All his possessions were scattered around the backpack—with the exception of his passport and an undetermined amount of American currency.

Many people close to the case believe that Weisfeiler was apprehended by a military patrol and handed over to Colonia Dignidad, an enclave of ultrarightist German immigrants founded and led by ex-Nazi Paul Schäfer.

Colonia Dignidad had been used by the Allende government for the deten-
tion and torture of dissidents and political opponents.

The U. S. State Department has taken an interest in Boris's case, and
there have been many articles about the incident (see, for example,
http://www.weisfeiler.com/boris/). The book *The Last Secret
of Colonia Dignidad* was written by Carlos Basso about the Weisfeiler incident. Some people, including Boris's sister and Harvard Mathematics
Professor David Kazhdan (1946–), believe that he is still alive.

There are many tales of Newton's (Isaac Newton, 1642–1727) absent-
mindedness. For him, his scientific thoughts were all-absorbing. He had
neither kith nor kin, and few responsibilities, when he was a Professor at
Cambridge. And he took full advantage of his position. His absorption also
applied to his religious devotion. A typical story is this:[1]

> On one occasion, when Newton was giving a dinner to some friends
> at the university, he left the table to get them a bottle of wine; but, on
> his way to the cellar, he fell into reflection, forgot his errand and his
> company, went to his chamber, put on his surplice, and proceeded to
> the chapel. Sometimes he would go into the street half dressed, and
> on discovering his condition, run back in great haste, much abashed.
> Often, while strolling in his garden, he would suddenly stop, and then
> run rapidly to his room, and begin to write, standing, on the first piece
> of paper that presented itself. Intending to dine in the public hall, he
> would go out in a brown study, take the wrong turn, walk a while, and
> then return to his room, having totally forgotten the dinner. Once hav-
> ing dismounted from his horse to lead him up a hill, the horse slipped
> his head out of the bridle; but Newton, oblivious, never discovered it
> till, on reaching a tollgate at the top of the hill, he turned to remount
> and perceived that the bridle which he held in his hand had no horse
> attached to it.

It is said of Isaac Newton that he thoroughly devoted himself, for a period
of 18 months or more, to writing the *Principia*. He essentially gave up all

---

[1]Taken from [MOR, Quote 1022].

earthly considerations during this period. He often forgot to eat his lunch. Or, if he had eaten it, he forgot that he had eaten it.

Complex analyst Peter W. Jones (1952–) began his mathematical career as a Dickson Instructor at the University of Chicago. He later rose through the ranks and became a full Professor there. He is now Professor of Mathematics at Yale University.

Anyway, Peter's first apartment in Chicago was a unit in university housing. It dated back to the dawn of the machine age, and offered a number of technological challenges. Ancient radiators battled newer cooling systems, drafty windows, and various heating and cooling vents. Peter told me that, in the dead of winter, it would actually *rain* in his living room.

J arrived as an Assistant Professor at UCLA in 1974. A number of strange initiations marked my advent at that fine institution.

The Chairman of the Math Department held a "welcome back" party at the beginning of the school year. At one point one of the senior faculty— David Cantor (1935–)—came up to me, introduced himself, and said, "You're a young man. You've got your whole life ahead of you. So let me give you some advice. If you are driving down the road and you are stopped by the cops, *don't get out of the car*. Don't make any sudden moves. Keep your hands visible at all times. Tell the policeman that you are going to reach for your wallet and then do so *very slowly*, keeping your hands visible. Be unfailingly polite." And so forth and so on. Seems that this was during the tenure of Daryl Gates as Chief of Police, and the LAPD had a certain reputation.

I was also cautioned to go easy on the Chairman. He had just been featured in the Santa Monica *Evening Outlook* for chasing his wife around the kitchen with a butcher knife. He had been under a lot of stress.

Soviet mathematician Gennadi M. Henkin (1942–) is one of the heroes of modern analysis. In his early career he studied concrete Banach spaces. He made a real splash around 1970 when he produced explicit integral formulas for the study of several complex variables. This work really changed the face of the subject. Henkin's interests have meanwhile gone on to scattering theory and Berger's equation and many other topics. But he is still remembered fondly for the impact he had on complex analysis.

Henkin suffered in Soviet Russia from anti-Semitic policies that were prevalent in the 1960s and 1970s and 1980s. He presently enjoys a fine position at the University of Paris. But for many years in Russia he was unable to obtain a proper job in a university mathematics department. His positions were primarily in various Economics Institutes and Economics Departments. And of course he was expected to produce work in Economics.

In fact he did so, and his Economics scholarship was creditable and worthwhile. But on his own time Henkin continued to study pure mathematics, and his results there were seminal. In fact Henkin reports that, when he published a good theorem in several complex variables, his salary at the Economics Institute was decreased.

Victor S. Miller (1947–) was a math graduate student at Harvard beginning in 1968, and is the source of this story. In those days Mathematics at Harvard was located in a grand old building at 2 Divinity Avenue in Cambridge. The graduate students were a rowdy bunch—many of them fresh from the turbulent political activities at Columbia University—and they were left largely to their own devices. The faculty seemed to be rather distant and forbidding. The students frequently groused about the inaccessibility of the faculty. In response to the student unrest, and sensitive to student agitation, in the Spring of 1969 the faculty met and passed a resolution saying, "All grad students must speak with a faculty member at least twice a year."

This was not a hollow effort. The following Fall, first-year graduate student Ken Ribet (1948– , now a well-known Professor at U. C. Berkeley and celebrated for his role in the proof of Fermat's Last Theorem) came up the stairs to the math office. Standing there was David Mumford—soon to be a Fields Medalist. Mumford greeted Ribet with a hearty handshake and said, "Welcome to Harvard. We've voted to talk to you students this year."

In the late 1960s, when Math at Harvard occupied 2 Divinity, the first floor was occupied by the Yenching Institute. The mathematicians and the Chinese spiritualists spent most of their time ignoring each other.

Nevertheless, with two active groups in a building of moderate size, the facility was bursting at the seams. So everyone was quite excited about the imminent construction of the new Science Center. This was to be the new home of mathematics.

Andrew Gleason was Department Chair at the time, and he uncharacteristically left his office door open each day so that people could come in and examine the three alternative plans for the Math Department facilities in the new Science Center. When graduate student Victor Miller went in to have a look, Fields Medalist Heisuke Hironaka was among the other interested parties. There was considerable discussion of the different choices, during which Miller offered, "2 Divinity is such a nice building. Why can't we (the mathematicians) just take over the whole building?" Hironaka then mumbled (just loud enough for those nearby to hear), "Yes, if we could only get rid of those Chinese ...."

Heisuke Hironaka

In 1973, Fischer Black (1938–1995) of Harvard and Myron Scholes (1941–) of Stanford revolutionized the world of finance by developing the first option-pricing scheme. One of the remarkable things about their work is that it is mathematically very sophisticated. In fact it uses stochastic integrals. Nowadays, when a professor in a mathematics department teaches a graduate course on measure theory, he or she can expect to see a number of finance and business students in the class. We can thank Black and Scholes for this development.

These two scholars were awarded the Nobel Prize in Economics for their joint work. This was big news at Stanford, and an article about their triumph appeared in one of the Stanford periodicals that is circulated to faculty, staff, and alumni. Unfortunately, some inept editor used a spell-checker on the article and he ended up changing every occurrence of "Myron Scholes" to "moron schools." And that is how it appeared in the published article.

Once Dennis Sullivan (1941–) was giving a math talk at U. C. San Diego to a rather sophisticated audience that includ-
ed Raoul Bott (1923–), Fields Medalist Michael Freedman (1951–), and others. At one point Sullivan announced that he was going to discuss "Ahlfors's theorem." He wrote **Proof:** on the board and began to explain the proof.

Bott asked Sullivan what the theorem said. The reply: "I thought I'd save time by not stating the theorem. You'll see what's going on as you follow the proof." After a while, everyone was quite frustrated. So Michael Freedman said, "Come on, Dennis. Tell us what the theorem says." Sullivan looked a bit shamefaced and said, "I forget."

Michael Freedman

Pierre-Simon Laplace (1749–1827) was one of the most significant mathematicians of his day. As a Professor at the École Militaire, he had the opportunity to be examiner of the young Napoléon Bonaparte. After the French Revolution, Napoléon appointed Laplace to be the Minister of the Interior of France's new cabinet. But the Emperor dismissed Laplace after only six weeks, saying that "Laplace saw no question from its true point of view; he sought subtleties everywhere, had only doubtful ideas, and finally carried the spirit of the infinitely small into administration." Nonetheless, Laplace's scientific reputation continued to grow. He was ultimately made a Marquis. He is now buried in Paris's famous Père Lachaise Cemetery, about a stone's throw from Jim Morrison's memorial.

Brown University, in the mid-to-late 1970s, had one of the strongest geometry/ algebraic geometry groups in the world. Robert MacPherson (1944–), Jean-Luc Brylinski (1951–), Paul Baum (1936–), Benedict Gross (1952–), William Fulton (1939–), and Joe Harris (1951–) all graced the halls of the Brown Math Building with their presence. They enjoyed vigorous mathematical activity and collaboration.

But one day Brown got a new Dean who decided that undergraduate education was really the thing. Of course Brown has always been famous for its undergraduate program, but this guy took it to the limit. He told the faculty that if they wanted to do research they would be doing it as a hobby. He would not support them financially or in any other way: not for colloquia, not for travel, not for anything. So these six guys decided to leave:

- Dick Gross is now a Professor at Harvard;
- Joe Harris is now a Professor at Harvard;
- Robert MacPherson is now a permanent Professor at the Institute for Advanced Study;
- William Fulton went on to be a Professor at the University of Chicago and is now at U. of Michigan;
- Jean-Luc Brylinski and Paul Baum are Professors at Penn State.

Nobody knows what has become of the Dean.

$\mathcal{B}$ill Browder (1934—) of Princeton University was slated to give a plenary talk at the International Congress of Mathematicians in Nice in 1970. Of course this is a huge deal, one of the great encomia of the mathematical life. The magic moment arrived, and an appointed individual gave a flowery introduction in which he praised Browder's many mathematical accomplishments, and all the important ideas he had created. Then—immediately—the power went off. No lights, no overhead projector, no electric blackboards, no nothing. The lecture was in fact never delivered.

Some wag came up to Browder afterwards and told him that it was the best talk of the meeting.

$\mathcal{E}$dmund Landau and G. H. Hardy had many common interests in analytic number theory. At one point, Hardy was invited to visit Landau in Germany. Of course Landau went to the train station to meet the eminent man. As Hardy stepped down from the coach, Landau rushed up to meet him: "Do you have a new method for treating the minor arcs (for the unit circle in the complex plane)?" Hardy looked irritated and said, "No. I am no longer interested in major and minor arcs." Landau was sorely taken aback, until he realized that the person to whom he spoke was not Hardy but one of the senior Göttingen students masquerading as Hardy. The genuine Hardy, with appropriate interests, was soon on hand.

*B*arry Mazur (1937—) is one of the statesmen of modern mathematics. An accomplished number theorist, Mazur's PhD thesis was in topology (under the direction of Ralph Fox, 1913—1973). In fact Barry once told me modestly at dinner that what he did in his PhD thesis was to prove the Generalized Schoenflies Conjecture. As the reader may know, this is the higher-dimensional version of the Jordan curve theorem. It is a *major* mathematical result. Mazur's 26-page PhD thesis, entitled *On embeddings of spheres*, is still legend.

Barry Mazur has always been a prodigy. He left the Bronx High School of Science after his junior year in order to go directly to MIT. He left MIT after his sophomore year to go to Princeton for graduate studies. Barry told me that he also left Princeton after one year to go study in England. To make a long story short, the only degree that Mazur ever got was the PhD from Princeton. This little fact evidently caused no end of headaches as Barry's career unfolded. When he would submit his NSF grant proposals, they would invariably be returned with a request for information about his high school diploma and college degree. The Administrative Assistant, Mary McQuillen, would write back and say, "This is not an omission, PhD is Professor Mazur's only degree."

The next story comes from Peter Markstein (1937–). MIT was impressed by Barry Mazur's sudden notoriety on proving the Schoenflies theorem. The venerable institution decided to award Barry his missing B.S. degree. But there was a catch. At the time (1958), it was required that all MIT undergraduates complete a course of Navy ROTC. The MIT administration tried to get the Navy to waive this requirement for Barry, but the Navy stuck to its guns. In the end, a rather gentlemanly compromise was negotiated. If Barry would complete one 2-hour drill successfully, then they would declare him to have passed the requirement. According to Peter Markstein (who witnessed the event), Mazur arrived 5 minutes late, dressed in Bermuda shorts and a T-shirt, wearing sunglasses and drinking a beer. Things went from bad to worse, and he was kicked out of the drill. So Barry Mazur still does not have a Bachelor's Degree from MIT.

On the other hand, Barry Mazur is now a University Professor at Harvard. This means that he no longer has a departmental affiliation. He is a Professor of the entire institution, and can teach courses in any department he likes. His book *Imagining Numbers* [MAZ] is an extraordinary polemic that communicates the subtleties of the complex number system to philosophers and social scientists. Barry Mazur continues to have a special and profound effect on all of us.

$\mathcal{M}$arshall Stone (1903–1989) is said to have run the University of Chicago mathematics department (in the 1940s) with a firm and imperious hand. One day student Bertram Kostant (1927–) was humbly and patiently waiting outside Stone's office for an opportunity to speak with the great man. André Weil came to the door, pushed his way through, and strode up to the desk to speak to Stone. Marshall Stone gave Weil a grave look and pointed out that the student had been there first. Kostant always held Stone in extremely high regard after that moment.

$\mathcal{I}$n the academic year 1988–1989, a special year in several complex variables was held at the Mittag-Leffler Institute in Djursholm, Sweden. This is the location of the grand old house of Gösta Mittag-Leffler (1846–1927). It now has housing for visitors and is a stunning venue for mathematical research. All the best people spent time at Mittag-Leffler during that year. One of the residents was a particularly distinguished Russian (who shall remain nameless, for reasons that will soon become clear). One day he decided to wash his clothes. But he was trenchantly unfamiliar with Western technology and conveniences. He went to the laundry room, tossed a quantity of powdered laundry detergent into the *dryer*, and then manually dumped several buckets of water into the dryer. Finally, he added his clothes and turned the contraption to "On." The result was smoke, sparks, boiled water, burned clothes, fried soap, and a general mess. It took the Mittag-Leffler staff a day and a half to clean up the catastrophe, and of course the dryer had to be replaced. But mathematics forged ahead at old Mittag-Leffler.

$\mathcal{P}$aul Erdős never liked the government, nor any form of authority. He was dreadfully upset with the state of affairs in the early 1950s in the United States when the McCarthy era took hold. In those days, the FBI kept a large file on Paul Erdős; they particularly did not like the fact that he corresponded with scientists in Hungary and China.

In 1954, Paul Erdős was invited to speak at the International Congress of Mathematicians (ICM) in Amsterdam. Of course this is a very great honor,

Paul Erdős

and Paul was eager to go. At the time the invitation arrived, Paul was visiting at the University of Notre Dame. The Immigration Service sent an officer to the Math Department there to grill him on his political leanings.

"What is your opinion of Marx?" queried the officer. "I do not feel competent to judge Marx," said Paul Erdős. "Because I have read only *The Communist Manifesto* by him. But I do believe he was a great philosopher."

Paul Erdős was then asked whether he would visit Hungary if the Hungarian authorities guaranteed that he could leave the country whenever he wished. "Of course I would!" he replied. "My mother lives there as well as some of my best friends."

Paul's visa application was denied. He decided to go to the ICM anyway. He determined to pack up his meager possessions and simply quit the United States—for an indeterminate period of time. His American friends begged him to stay, to wait a year and submit another visa application. Paul was intransigent. He spent his last night in the U.S. with Harold Shapiro (1928–), who begged him not to go. Shapiro was quite aggressive, and very critical of Paul Erdős's decision: "I should knock you on the head and tie you up to stop you from leaving!" cried Harold. But Paul Erdős remained steadfast. "O.K., then tie me up!" was his reply. His case was reconsidered many years later in 1963.

Paul Erdős had analogous problems with the Hungarian authorities in 1973. As a result, he was unable to return to Hungary (his native land!) until 1976.

$\mathcal{T}$hose who hung out with Erdős were accustomed to the activity of "Uncle Paul Sitting." Paul was usually quite helpless with most things that were not mathematics. This was perhaps a product of his upbringing. His parents were overly protective; indeed, Paul did not butter his own bread until, at the age of 21, he was on a trip to England. When he visited a university, he expected people to fetch envelopes and tablets and pens for him, make phone calls, give him rides, wash his clothes, feed him, and take care of all his other needs. Erdős did seem, at one point in his life, to have an amorous liaison—with a woman. After a time it became apparent to her that Paul really valued her because she gave him rides to wherever he needed to go. End of relationship.

Erdős could remember two dozen phone numbers at a glance, but appeared to be absent-minded about many things. Once, when visiting Cal Tech, he lost his sweater twice in the same day. The first time it was recovered and the second time not. When he visited my own university we asked him for his Social Security number so that we could pay him. He did not know it. At dinner, since no mathematics was being discussed, he fell asleep.

$\mathcal{U}$rban Cegrell (1943—) is an active and influential mathematician at the University of Umeå in Sweden. The department is a fairly happy and contented place, and the home of a lot of good mathematics. Several years ago, according to Cegrell, there was some discontent. Seems that the university administration came to the Mathematics Department and informed its denizens that it wanted to rename the group the "Department of Data-Logic."

$\mathcal{A}$round 1940, William Fogg ("Foggy") Osgood's calculus text was still used at Harvard. One remarkable feature of this book is that Osgood rejected limits as "a bit much" for Harvard freshmen. He favored infinitesimals. Leonidas Alaoglu, of Banach-Alaoglu theorem fame, was set to teach calculus that year. He did not necessarily approve of Foggy's approach to the subject. He is quoted as saying to his class, "Gentlemen, we now come to Chapter VII on infinitesimals. Please take pages 123–150 between thumb and finger of the right hand, and tear the pages from the book." And so it was done.

$\mathcal{D}$esmond MacHale (1946–) is a well-known mathematical wit, author of the book [MACH]. He has reported in this way on his experience at a colloquium:

> Recently, I attended a mathematical lecture given by a guest speaker where absolutely nobody, except possibly the speaker, had the remotest idea what was going on.... After the speaker had finished over an hour later to an enthusiastic round of applause, the chairman asked for questions and, of course, there was a deathly and highly embarrassing silence. Then and there I resolved to put together a collection of universal questions for use in such situations.... The following is the list I came up with.
>
> 1. Can you produce a series of counterexamples to show that if any of the conditions of the main theorem are dropped or weakened, then the theorem no longer holds?
> 2. What inadequacies of the classical treatment of this subject are now becoming obvious?
> 3. Can your results be unified and generalized by expressing them in the language of Category Theory?
> 4. Isn't there a suggestion of Theorem 3 in an early paper of Gauss?
> 5. Isn't the constant 4.15 in Theorem 2 suspiciously close to $4\pi/3$?
> 6. I'm not sure I understand the proof of Lemma 3—could you outline it for us again?
> 7. Are you familiar with a joint paper of Besovik and Bombialdi which might explain why the converse of Theorem 5 is false without further assumptions?
> 8. Why not get a graduate student to perform the horrendous calculations mentioned in Theorem 1 in the case $n = 4$?
> 9. Could you draw us a simple diagram to show what the situation looks like for $n = 2$?
> 10. What textbook would you recommend for someone who wishes to get students interested in this area?

11. When can we expect your definitive textbook on this subject?

12. Why do you think there was such a flurry of activity in this area around the turn of the century and then nothing until your paper of 1979?

13. What are the applications of these results?

And if that won't stop most any speaker cold, then I don't know what will.

$\mathcal{B}$ernie Shiffman (1942–) is a Professor at the Johns Hopkins University, and an excellent geometer. One day, some years ago, he went to the open house at his son's high school. Each teacher put on a dog-and-pony show, and Bernie dutifully went to every room to sample what the teachers had to offer. He was a bit surprised when he entered the math teacher's room, because she had various triangles exhibited on the blackboard that illustrated ideas from trigonometry. And certain angles were shown to have sine equal to 3 and cosine equal to 4.5. After the presentation, Bernie went up to talk to the teacher about it. He is a humble sort of guy, so he said, "You know, it's been a long time since I've had any trig. But I always thought that sine and cosine took values between –1 and 1." The teacher smiled. "Oh no," she said. "You just make the triangle larger."

$\mathcal{G}$. H. Hardy's passion for mathematics was legendary. There was hardly anything else—except possibly for the game of cricket—about which he cared so fervently. As a Fellow at Cambridge, he had all his worldly needs attended to—housekeeping, food, etc.—so he could concentrate all his powers and all his attention on his scholarly work. As a result, the man was acutely unworldly. In particular, Hardy had a morbid suspicion of mechanical gadgets. He never used a watch. And he was particularly leery of the telephone. In his Trinity rooms, or in his flat on St. George's Square, he was heard to say (in a slightly sinister and certainly very disapproving tone), "If you *fancy yourself* at the telephone, there is one in the next room."

$\mathcal{J}$oseph F. Ritt (1893—1951, after whom the Columbia University endowed postdocs are named) was the standout mathematician at Columbia during the 1930s and 1940s. He was a highly original and inward-looking thinker who developed his ideas by reading the great classics: Jacobi, Abel, Liouville, and others. He openly rejected many of the modern trends, including Lebesgue's measure theory. In some instances his contempt for modern frippery knew no bounds. In particular, as a longtime worker using only real or complex numbers, he referred to finite fields as "monkey fields."

J. F. Ritt used to frequently lament that, despite his many accomplishments and distinctions, he had been awarded none of the major national mathematics prizes. Sometimes he would joke about the matter. An epitaph that he composed for his gravestone read

> Here at your feet J. F. Ritt lies;
> He never won the Bôcher prize.

*O*ne of Erdős's favorite problems grew out of Ramsey theory. It asks, "Let k be a positive integer. How many people $N = N(k)$ do you need in a room to be sure that at least k people are mutually acquainted or at least k people are mutually unacquainted?" When $k = 2$, the answer is trivially $N = 3$. When $k = 3$, the answer is 6 but this is not entirely obvious. Here is a proof:

Call one of the 6 people Igor. Of the 5 remaining people, Igor either knows 3 of them or doesn't know 3 of them. Say the first option obtains. So say that Igor knows A, B, and C. If A, B, C are mutually unacquainted then we are done. Otherwise two of them are acquainted, and both are acquainted with Igor, so again we are done.

We leave it to the reader to verify that $N = 5$ will not do when $k = 3$ (think in terms of a pentagon).

For $k = 4$, the answer is $N = 18$ and that is quite difficult to see. For $k = 5$, it is known that $42 \leq N(k) \leq 55$. It has required 40 years to obtain these estimates. And we are far from knowing the exact answer. For $k = 6$ we know that $69 \leq N(k) \leq 102$, but determining the exact answer is considered to be slightly worse than hopeless. Erdős put the matter this way:

If an evil spirit would appear and say, "Tell me the value of N when k equals 5, or I will exterminate the human race," it would be best to get all the computers in the world to try to solve the problem. But if the evil spirit would ask for the value when k equals 6, it would be best to try to exterminate the evil spirit.

He goes on to say, "And if we could get the right answer just by thinking, we wouldn't have to be afraid of him [the evil spirit], because we would be so clever that he couldn't do us any harm."

Reading Weil's memoir [WEI], one gets the sense that the young Weil felt that the Bourbaki group was in effect re-inventing mathematics for the twentieth century. In particular, they endeavored to standardize much of the terminology and the notation. Weil in fact takes credit for inventing the notation $\emptyset$ for the empty set. The way he tells it, the group was looking for a good way to denote this special set; Weil was the only person in the group who knew the Norwegian alphabet, and he suggested that they co-opt this particular letter.

Emmy Noether (1882–1935) suffered the prejudices against women that were common in her day. But she was such an enormous talent that she had many admirers. Certainly David Hilbert (1862–1943) was one of her staunchest defenders. In 1919 he was arguing for her qualification as a faculty member in Göttingen. His strongest opponents were the philologists and the historians. At one point he addressed the council of the university with these words: "I do not see why the sex of the candidate should be an argument against her appointment as Privatdozent; after all, we are not a bath-house …"

Emmy Noether

With the free and easy nature of modern society, it is sometimes difficult to feature some of the nuances of behavior in a bygone age. For example, in the Victorian era it was considered extremely rude either to smile or to laugh. J. C. Burkill (1900–1993) tells us that

> Littlewood belonged to a generation before Christian names supplanted surnames, and so *a fortiori* did Hardy. Of all Littlewood's relationships with men, that with Hardy must have been one of the closest. It is inconceivable that either would have used the Christian name of the other. Their letters would begin "Dear H" and "Dear L," if there was an allocution at all. A typical letter in 1920 began "Quite a rush of ideas this morning ...."
>
> With the later widespread use of Christian names, perhaps four or five of more than a hundred Fellows of Trinity called him Jack (without a response in kind?); only the latest of his five Masters could speak *ex cathedra* of Jack Littlewood.

As noted above, G. H. Hardy (1877–1947) and J. E. Littlewood (1885–1977)—although they were the closest of friends—adhered to a rather formal, quasi-Victorian social style. They showed occasional bursts of wit—of a rather stuffy, scholarly sort. One example is this comment by Littlewood about Hardy's contribution to the notation adherent to Dedekind cuts:

> The letters L, R (in a Dedekind section) for which a generation of students is rightly grateful, were introduced by me.
>
> In the first edition of *Pure Mathematics* [Hardy's famous volume] they are T, U. The latest editions have handsome references to me, but when I told Hardy he should acknowledge this contribution (which he had forgotten) he refused on the ground that it would be insulting to mention anything so minor. (The familiar response of the oppressor: what the victim wants is not in his own best interests.)

Gian-Carlo Rota (1932–1999) was an energetic and broadly trained mathematician who spent most of his career at MIT. He was a student of Jacob Schwartz (1930–) at Yale, and cut his teeth doing functional analysis. In fact he was one of the many student assistants in the preparation of the monumental work *Linear Operators* by Dunford and Schwartz. Rota later gravitated to discrete mathematics and became one of the leading lights of partition theory, Young diagrams, and other parts of combinatorics.

Rota worked with an intensity and passion unknown to most of us. If he had an important lecture to prepare, he would isolate himself for several days, and not sleep nor eat. When in his sixties, he developed the habit of eating only one meal per day and of ingesting twelve tablets of Ritalin daily. Unfortunately, this was his undoing. When Rota was only sixty-seven years old, his heart stopped—he died in his sleep.

*O*ne of the great things that Rota did in the last years of his life was to assemble lists of precepts that mathematicians should live by. One of my favorites of these is "Ten Rules for the Survival of a Mathematics Department". Part of Rota's motivation here was pure good will, and part of it was his perception of a need to defend ourselves against such upstart subjects (which are stealing all our grant funding) as artificial intelligence and chaos theory. I reproduce the rules here, without the rather lengthy explanation that Gian-Carlo provided for each.

1. Never wash your dirty linen in public.
2. Never go above the head of your department.
3. Never compare fields.
4. Remember that the grocery bill is a piece of mathematics, too.
5. Do not look down on good teachers.
6. Write expository papers.
7. Do not show your questioners to the door.
8. View the mathematical community as a united front.
9. Attack flakiness.
10. Learn when to withdraw.

*W*hen Gian-Carlo Rota ran the journal *Advances in Mathematics* he did so with an iron fist. He kept all the records in his head, and was quite autonomous in deciding what would be published and what not. He wrote virtually all the book reviews himself. I thought that he had done a lovely job of reviewing Paul Halmos's memoir *I Want to Be a Mathematician*. He made a point of emphasizing what was good about the book and glossing over some of the rough spots. It was a really gentlemanly job. And I wrote Rota a letter telling him how much I appreciated his effort. He replied with a pre-printed postcard that said, "Keep those cards and letters coming, folks."

I must say that I found this quite irritating, but fate offered me a slight opportunity for a comeback. Some years later Rota and I became pretty friendly and he used to send me drafts of articles that he had written. One such was about "Ten Commandments for the Mathematician." He asked for my criticism, and whether I would add any commandments. Well, I really liked what Gian-Carlo had written and told him so. But I did suggest an eleventh commandment:

If a serious mathematician sends you an honest and well-thought-out letter, then you should not reply with a pre-printed postcard.

*O*ne afternoon years ago, Paul Halmos drove Johnny von Neumann to his house in Princeton. There was to be a party there later that evening, but Halmos didn't trust himself to be able to find the house again (alone). So he asked Johnny how he'd be able to recognize the house. The reply was, "That's easy. It's the one with that pigeon sitting by the curb."

*J*ohn von Neumann liked to drive, but he was not particularly good at it. There was a "von Neumann's corner" in Princeton at which his autos repeatedly had trouble of one sort or another. He offered this explanation for one particular crack-up: "I was proceeding down the road. The trees on the right were passing me in orderly fashion at 60 miles per hour. Suddenly one of them stepped in my path. Boom!"

*J*ohn von Neumann gave many parties at his house. He enjoyed drinking, though he was not a heavy drinker. He gave boisterous parties—with ample alcohol provided—in Fine Hall, and there are tales of empty liquor bottles sailing out the windows as the parties progressed. In a roadside restaurant von Neumann once ordered a brandy with a hamburger chaser.

*I*n the 1920s, the Columbia University mathematics department was in something of a slump, and the administration was anxious to help the group build back up to its proper stature. The department, in effect, made an ordered list of the world's greatest mathematicians and began making them offers. The first was Hermann Weyl, and he was tendered a princely salary and benefits. Many special conditions and arrangements were made in order to please, and to entice, Weyl to Manhattan. One of these perks was that Lulu Hoffman, Weyl's young assistant, was to come with Weyl. This was a tricky point, because Columbia had only male professors in those days. But no matter, it was arranged for her to teach at Barnard College. Hoffman was actually to be the first female math teacher at Barnard. The bargaining with Weyl continued for a considerable period, and in the end (as we all know) Weyl declined the offer. In his letter turning Columbia down, Weyl noted that Göttingen was the center of the mathematical universe, that he was quite happy there, and that he did not wish to change things by accepting Columbia's offer. The young instructors at Columbia used to joke about the vision of Hermann Weyl, on a deck chair of the steamship S. S. Bremen, crossing the ocean to New York with the center of gravity of world mathematics following obediently some one hundred yards behind the propeller's wash.

*J*ohn L. Kelley—author of the wonderful book *General Topology* which was mastered by several generations of mathematicians—grew up in poverty during the depression. He barely scraped through college and had little hope (from a financial point of view) of going to graduate school. So he determined to be a high school teacher. In his first semester at UCLA he enrolled for three courses to prepare him for a secondary school credential. In his words, "The courses were pretty bad and besides, the grading was unfair, e.g., I wrote a term paper for Philosophy of Education and got a *B* on it; my friend Wes Hicks (1910–), whose handwriting was better than mine, copied the paper the next term and got a *B*+, and our friend Dick Gorman (1913–) *typed* the paper the following term and got an *A*."

Dick Gorman has a slightly different memory of his relationship with Hicks and Kelley:

> The three of us knew each other from L.A.J.C. [Los Angeles Junior College] when we were all majoring in math. We took a course in Education at UCLA together. For the final exam we sat in this order Gorman, Hicks, Kelley. Hicks could see both papers and he followed the following procedure. If Kelley and I had the same answer, he used that answer for his paper. If we differed he used his best judgment, which meant he probably used Kelley's answer. In any case he

couldn't help but [pass] as Kelley and I were later chosen to be members of Phi Beta Kappa in April 1936.

*J*ohn L. Kelley was a talented mathematician, but he struggled with uncertainty and frustration as we all do. As he tells it,

> The mathematicians then (1935–1945) were like mathematicians now, only more so. John Wehausen (1913–), an early editor of the *Mathematical Reviews*, once told me that mathematics was one of the psychologically hazardous professions. "Every mathematician, for most of his early life, is the brightest person he knows, and it's a great shock when he finds there are people that can do easily things that are very hard for him" according to John. I think that this is true, and that within every mathematician, more or less suppressed or laughed at, is an arrogant little know-it-all, and simultaneously a stricken child who has been found wanting. Johnny von Neumann has said that he will be forgotten while Kurt Gödel (1906–1978) is remembered with Pythagoras, but the rest of us viewed Johnny with awe.

*R*ichard Feynman, Nobel-Laureate physicist, is a favorite among mathematicians. Even though he liked to make fun of mathematicians, he was in many respects one of us.

Feynman spoke aggressively against paranormal events. He really thought that people who espoused that stuff were pretty silly, and he made no bones about it. "But," he said, "I did have this one rather striking experience. One day in college, I was studying hard in my room, really concentrating on my work. And I suddenly got this vision of my dear, sweet grandmother—dead in her casket. I was dreadfully upset. I ran to the phone and dialed the number for our house—grandma had lived with us at the time." At this point Feynman gave a meaningful pause to his story. "Granny answered the phone. She was fine."

*I*t is all that most of us can do to earn a PhD in mathematics and then make our way in the profession. But there are exceptional individuals among us.

Schlomo Sternberg of Harvard is a Doctor of Mathematics and also a Doctor of Rabbinical Studies. Murray Gerstenhaber of Penn has a PhD in math but he also holds a law degree. Richard Gundy, a distinguished harmonic analyst, holds degrees in Statistics and Psychology but none in Mathematics.

*J*n the early 1950s, Columbia University was home to a number of distinguished mathematicians: Claude Chevalley (1909–1984, who is said to have refused admission into his linear algebra course to anyone who had previously studied matrix theory), Harish-Chandra (1923–1983), and others. There were a number of French visitors, including Jacques Hadamard (1865–1963) and Arnaud Denjoy (1884–1974).

At one point, Denjoy was giving a lecture series to an audience that began with a substantial number of subscribers but quickly frittered down to just three students. After a few more lectures, these three decided to go on strike, claiming that their situation was untenable. The entire department was in an uproar. Finally the strikers, after much urging, agreed to return to the lecture hall but on one condition: that Denjoy should cease lecturing in English and instead lecture in French.

*J*n the Summer of 1977 a number of us young analysts were hanging around UCLA trying to prove great theorems. I particularly remember Mike Taylor (1946–) and Bob Jensen (1949–)—who had quite divergent styles. Mike would have a pot of coffee and a carton of cigarettes on his desk and would spend all day chain smoking and chugging coffee. When he and I would discuss Lipschitz spaces, he would be bouncing off the walls. Jensen, on the other hand, had a pitcher of *Goofy Grape* on his desk. He was much more mellow.

*J*t has been recorded elsewhere that Ron Graham (1935–) put great effort into caring for Paul Erdős in his final years. Graham even built a wing onto his house so that Erdős would have a place to live. Graham kept track of Paul's bank accounts and correspondence. But Daniel Kleitman of MIT did Erdős's income taxes. Everyone, it seems, pitched in to help Erdős.

*W*hen I was a graduate student at Princeton, the mail was always erratic. There was a period of time when virtually anything coming into the Math Department through the U.S. mail was tagged as "damaged in transit". Usually, if there was any damage at all, it was minimal.

The whole matter became a departmental joke. Each day we would sit in the coffee room (known as the "Commons Room") and swap tales about our damaged mail. But one day one of us got a manila envelope, tagged as usual as a damaged item. He opened it up and inside were only ashes.

*I*n the late 1960s, Tony Tromba (1943–) was a hard working young mathematician on leave at the University of Pisa—one of the most distinguished mathematics departments in all of Italy. Now Tony's tenure decision was a few years off, and he was putting in all his effort to create new mathematics and publish excellent papers. It was of course the custom in Italy in those days for professors to take long, languorous lunches and then to enjoy a siesta afterward. Not Tony. He would scarf down a piece of pizza and immediately resume his work.

This habit of Tromba's rather annoyed the Italians. They felt that he was not getting with the program. So they attempted a variety of devices to discourage Tony's work ethic. First they tried locking the library. This did not work, for he would simply get the materials he needed before the lunch break. Then they locked all the seminar rooms and offices, so that Tony had no place to work. This did not bother him; the weather was pleasant and he could sit in the square. What finally did the trick is that they locked all the bathrooms. Then Tony had to go home.

*I*n today's world we go through periods of tough times on the job market. Many a young mathematician ends up taking a job at a school of much less distinction than he/she ever anticipated. The lucky ones can battle their way back up the food chain to a good job at a good university. Others spend their careers doing more teaching and less research than they ever expected.

We sometimes forget that things were much tougher seventy years ago. Nathan Jacobson (1910–1999) relates

> …this was in the depth of the Great Depression. Salaries declined in some instances and there were very few new positions. Moreover, for the new Jewish PhDs the situation was further aggravated by anti-

Semitism that was prevalent, especially in the top universities—the only ones that had any interest in fostering research.

Ivan Niven (1915–1999) augments these remarks:

Although the early thirties were apparently tougher than the late thirties, the depression was not really over for mathematicians until 1943 or so. Paul Halmos, well-known as an eminent mathematician and a great expository writer and lecturer, with a PhD from Illinois in 1938, sent out over 100 letters that spring inquiring about possible openings, with no luck at all. [It might be added that the eminent Saunders Mac Lane very nearly accepted a job at the prep school Exeter, but was rescued at the last moment.] As he [Halmos] recounts on page 80 of his autobiography "..., the University of Illinois kept him on for a year at a salary of $1800 and a teaching load of 15 hours per week."

Another example of hardship is that of the late Thurman S. Peterson. He was unemployed in the fall of 1938, living with his parents in Los Angeles, when an opportunity arose for him at the University of Oregon because of a sudden, debilitating illness of the department head. Peterson and I were colleagues at Oregon for many years. Here's the rest of the story: after his undergraduate years at Cal Tech and a PhD at Ohio State in 1930, Peterson held a temporary post at the University of Michigan for two years, and then got a stipend at the Institute for Advanced Study. After two years there, he was slated to be an assistant to one of the senior members of the Institute in 1934–1935. However, in the spring of '34 an inquiry came in to the Institute about a suitable candidate for a teaching post at a private high school for girls in the Philadelphia area. Teaching mathematics to high school students did not have great appeal to Peterson, but he was urged strongly by the Director of the Institute to apply for the post, so as to open up a stipend there for some mathematician in need. Accordingly, Peterson did apply, and after an interview was awarded the position. Years later, he told me that four years of that work was all that he could stand, getting farther and farther away from university life. In 1938, Peterson quit, with no other employment in sight, and that takes us back to his unemployed status in Los Angeles in the fall of 1938.

(Laughlin) Andrew Campbell (1942–) was a graduate student at Princeton. According to legend, he needed to pass the German language exam, so he knocked on the door of the German examiner—that year it was Salomon Bochner (1899–1982).

He made the mistake of telling Bochner, a native German, that he had lived eight years in Germany. Bochner said, "OK, your exam is to give me eight meanings for the word 'um'—one for each year that you lived in Deutschland." The student passed, but not without a struggle. He said later that he wished he had lived fewer years in Germany.

$\mathcal{G}$uido (1928–) and Mary (1930–1966) Weiss were both talented young mathematicians who earned their PhDs under the direction of Antoni Zygmund in the late 1950s. They were one of the first "two body problems". And in those days the situation was nearly untenable. One university offered the two of them a single position that "they could divide up any way they wanted." The only school that would offer them both *bona fide* jobs was Washington University in St. Louis. Guido has been here ever since. Mary, unfortunately, died tragically a few years after she joined the faculty at Washington University.

$\mathcal{S}$teve Brady (1941–) is a mathematician at Wichita State University. His specialty is numerical analysis. At some point, about thirty years ago, the Wichita police determined that the city was harboring a serial killer—several murders had been linked in a manner that pointed decisively to this conclusion. In a correspondence with police the killer named himself BTK (for Bind them, Torture them, and Kill them). The police ended up retaining Brady initially to train detectives to use computers, later to create a database of information about the crime, and finally to become a full partner in the task force hunting the killer. For Brady this began as temporary assistance, but it ended up as an obsession. He remained a full-time employee in the Math Department at Wichita State but spent much of his time (outside university activities) for several years tracking the killer without pay.

The search involved studying crime scene evidence, pursuing clues from correspondence between police and the killer, trying to confirm or refute profiling information provided by the FBI, and looking for patterns linking the crimes and/or the victims' lives to the killer. Mathematically, it was a huge inverse problem of organizing the evidence to attempt to learn enough of the killer's habits, methods, personality, mental capacity, etc. to narrow the search to one man.

Brady received a Medal of Meritorious Service from the city for his dedicated efforts—the only non-police person ever to receive such an award.

During the 1990s, and until recently, correspondence from the killer and murders that could be tied to him with certainty stopped. Nobody knew why. Many believed the killer to be dead. The search and case went cold and, although murder cases remain open until solved, the case for most citizens faded into history. However, for police officers (some now retired) who worked the case and Brady the pursuit slowed in intensity but was still being worked on unofficially. More recently, in 2004, new evidence of the killer surfaced. Combining the data that Brady and the police team had amassed from years before with new evidence from the killer himself, and with DNA evidence from the killer's daughter, the police finally captured the culprit in February of 2005. Like the hit television show *Numb3rs*, this case of the serial killer BTK is a powerful instance of mathematics in action.

*S*teve Brady is remarkable in other ways. He enjoys taking exotic trips. Once, in 1961, Brady went hiking in the eastern part of Germany. This was an extensive sojourn, and took the better part of a month. On his return through Berlin, he found that the Berlin wall had been erected and he was suddenly surrounded by German soldiers, all with machine guns pointed straight at him. It was difficult, with his halting German, for Steve to talk his way out of that situation.

These days Brady organizes exotic safaris in Africa. These are elaborate trips in which all the food has to be flown in, native bearers and guides are engaged, and one is quite isolated from the trappings of civilization.

*C*raig Evans (1949–) is Professor of Mathematics at U. C. Berkeley. One semester he was assigned to teach a *very* large calculus class—perhaps 500 students. Now Craig is a good teacher and a real survivor. He rallied his resources, got his act together, and jumped into the task. One day, after about six weeks of struggling with this Herculean labor, he walked to lunch with a friend. "This class is rather large and impersonal," said Craig. "It is a struggle to teach it effectively and well. My goal this semester is to learn the names of all my T.A.s."

*S*ome mathematicians in Southern California were sitting around one day bemoaning their lousy working conditions. One savant complained that the bathroom on his floor had no hot water. Another griped that the classrooms never had an adequate supply of chalk. Yet a third pointed out that the walls of his office were so thin that he felt as though he had no privacy. John Bachar (1928–) of Cal State Long Beach trumped them all when he pointed out that, whenever his telephone rang, cockroaches ran out from under the phone.

*M*athematician Robert A. Bonic (1932–1990)—who made his first appearance in *Mathematical Apocrypha*—smoked marijuana one night in 1969 and had a vision. Like Kronecker, he deduced that God made the integers and that all else was derivative material concocted by man. He decided therefore to become a computer scientist.

Robert A. Bonic

Using connections he had, Bob managed to land an Assistant Professorship in Computer Science at the Courant Institute in New York City. There was just one catch. Bob had been a tenured full Professor of Mathematics at Northeastern University. He resigned that lofty position in order to move to Courant. He had grown accustomed, since he had a high-paying sinecure at Northeastern, to fight and argue with everyone all the time about everything. Bob continued this tradition as an Assistant Professor at Courant. The unfortunate upshot was that he was fired. So Bob started a new life.

Bob Bonic had many interests. One of his avocations was playing darts, and he was so good at it that he was banned from most of the dart bars in New York City. He ended up opening his own dart bar in Manhattan. His partner in business was Phil Ochs (1940–1976), the famous folk singer. Bonic's bar became quite the hangout: actor Robert De Niro and punk rock singer Patti Smyth used to spend time there.

Phil Ochs was manic depressive, and he ended up committing suicide. Bonic was the last person to see him alive before he went off and hanged himself.

Ultimately Bonic was run out of business because the Cosa Nostra wanted his space. He became an itinerant mathematician, traveling from university to university and doing a variety of jobs for different math departments— some of them quite essential (like supervising the calculus curriculum)—in order to keep going. He died at a young age of a brain tumor.

*W*hen Tony Tromba was a sophomore at Cornell, he was one day in the office of Alex Rosenberg, discussing issues arising from their linear algebra class. Tony happened to glance out the window, and he saw someone scaling up the side of the building. It was Bob Bonic! Presently Bonic reached the window, Tony stepped aside, and Bonic hauled himself into Rosenberg's office. He gave the two a "by your leave" and sauntered out the door.

Utter Utterances

Solomon Lefschetz (1884–1972) was famous for his tart tongue. Some say it was all an affectation with Lefschetz, and others claim it was the real thing. These two particular incidents come from eye-witness Carl E. Langenhop (1922–).

In the early part of 1949, H. Bode (1905–1982, head of the mathematics section of Bell Labs at the time) was scheduled to give a public lecture on the Princeton campus. The declared topic was cybernetics—a recently concocted notion due to Norbert Wiener. Lefschetz, Al Farnell (1917–1994), and Langenhop attended. At the end of the lecture was a question-and-answer session. One man stood up, ostensibly to pose a question; but he instead delivered a twenty-minute harangue about his views of cybernetics. When he finally took a breath, Bode made a few quick remarks and closed the session. Afterward Lefschetz said, "Princeton is full of horse's asses."

A few years later there was a conference on differential equations at Notre Dame. Lefschetz was there, paying careful attention to the talks and offering comments from time to time. One talk in particular was by an engineer and included quite a lot of elementary detail—the sort of thing that a mathematician would usually leave as "an exercise for the reader." After he had had quite enough of this trivial nonsense, Lefschetz spoke up and said, "Do you postulate that all engineers are blockheads?" [It should be remembered that Lefschetz began his scientific career as an engineer.]

Mathematician Bruce Palka (1943–) likes to quote the *Math Reviews* item that says that "Theorem 1, by its absence, would lend stature to this paper."

*O*ne day Jacob Bronowski (1908–1974) phoned up John von Neumann in the middle of the night to admit that von Neumann had been correct the day before in some discussions they had been having. The great von Neumann's grumpy reply was, "Call me only when I am wrong."

*I*n von Neumann's day, fellow mathematicians and physicists marveled at the speed with which he could analyze and solve complex problems. "Most mathematicians prove what they can, von Neumann proves what he wants," was a popular saying among mathematicians of the time.

*C*handler Davis (1926–) tells this story (and then casts doubt upon same) about John von Neumann. According to Chan, one could tell Johnny one's latest results, and the great man would listen intently. A typical response would then be, "Ya-ya-ya, that is obvious." Alternatively, he could respond, "But that's not true!" If the interlocutor was extremely lucky, if in fact what he/she proved was really new, and unknown to von Neumann, and not completely obvious to him, then he might respond with, "But that's not true!" and then (after further explanation was supplied) switch to, "Ya-ya-ya, that is obvious."

*A*mong the women who suffered from misogynistic prejudice in the first part of the twentieth century was Olga Taussky Todd (1906–1995). During World War II, she spent some time at Bryn Mawr College (home of Emmy Noether, 1882–1935) and then moved to Girton College, Cambridge— where she stayed for the duration of the conflict. During her job interview at Cambridge, one member of the committee asked, "I see you have written several joint papers. Were you the senior or the junior author?" Another member of the committee was G. H. Hardy. He exclaimed, "That is a most improper question. Do not answer it!" At another interview Todd was asked, "I see you have collaborated with some men, but with no women. Why?" Olga retorted that that was why she was applying for a position in a women's college.

Olga Todd was lucky enough to attend Emmy Noether's seminar for a few years. During that time, she wrote a parody of a well-known verse of the poet Wilhelm Busch. Loosely translated, it reads thus:

> There Olga waits before her classes,
> she trembles much and thinks of Hasse's.
> Miss Emmy comes, she's drawing nigh,
> with booming voice and glowing eye.

> Upstairs she climbs, she's almost here,
> Poor Olga waits in helpless fear.
> Thus Olga thinks: the die is cast,
> my chances to survive are past,

> I won't waste on her theory
> what little time remains to me,
> but blithely calculate away.
> Plucky of Olga, I would say.

The great philosopher and mathematician Willard Van Orman Quine (1908–2000) had all his work typed on a particular 1927 vintage Remington typewriter for which the "1" , "!", and "?" keys had been removed and replaced with specialized mathematical symbols. A student once asked him how he managed to write without using a question mark. Quine replied, "Well, you see, I deal in certainties."

Cornell is one of our better universities, located in Ithaca, New York. One of the notable features of life at Cornell is that Ithaca is rather isolated. It is almost cut off from civilization during the winter, and not easy to get to even during the summer. Mark Kac (1914–1984) used to say that, "Cornell is a centrally isolated university."

When celebrated Polish topologist Kazimierz Kuratowski (1896–1980) visited Washington University, fellow Pole Henry Schaerf (1907–) took him under his wing. Schaerf squired the great man everywhere, and at Kuratowski's talk Schaerf acted as a facilitator. Schaerf had invited people from all over campus. So the room was packed, and Schaerf was in his glory.

During the question-answer session following the formal presentation, a man from the Economics Department formulated a succinct query. It took less than one minute to pose. Schaerf assumed his facilitator role and took it upon himself to *interpret* the question for Kuratowski, taking several minutes of highly convoluted prose to do so. At the end Schaerf said, "This question is too complicated to make any sense. Next question?"

*T*he 1936 book *Functions of Several Complex Variables* by S. Bochner and W. T. Martin is a watershed work in the function theory of *n* dimensions. For a time it was virtually the only place to go if you wanted to learn the subject. At some point the book was translated into Russian. Then some good Samaritan, not knowing that an English version already existed, used language translation software to translate the book *from Russian* back into English. The title came out

Functions of Many Complicated Changeables

Copies of this work can still be found on some shelves.

A finance professor and a "normal person" go for a walk, and the normal person sees a $100 bill lying on the street. Of course the normal person wants to pick it up, but the finance professor says, "Don't try to do that. It is absolutely impossible that there is a $100 bill lying on the street. Indeed, if it were lying on the street, someone else would already have picked it up."

Another instance of the insidious effects of the Black/Scholes theory.

*C*al Moore (1936–) of U. C. Berkeley is a gifted mathematician—of the von Neumann algebra ilk—who about twenty years ago became involved in university administration. He held various high-level positions on the U. C. campus, and most recently has been chairman of the Mathematics Department. He has played a decisive role in enhancing the role of the Mathematical Sciences Research Institute in Berkeley.

Moore has a particular passion for mathematics teaching. He is frequently quoted as saying that current high school mathematics consists of a little 2000-year-old geometry, a little 800-year-old algebra, and some 200-year-old analysis. But my favorite Cal Moore quote, to which I subscribe absolutely, is this:

> When you teach the kids, tell them the truth. Tell them nothing but the truth. But, for God's sake, don't tell them the whole truth.

This wisdom is reminiscent of the thoughts of the short-story-writer Saki (HH Munro): "A slight inaccuracy can save hours of explanation."

*G*race M. Hopper (1906–1992) was a Rear Admiral in the Navy and one of the real pioneers of computer science. She coined the word "bug" to describe any artifact that causes trouble in a computer system. She tells the story as follows:

> Things were going badly. There was something wrong on one of the circuits. Finally, someone located the trouble spot, and, using ordinary tweezers, removed the problem—a two-inch long moth. From then on, when anything went wrong with a computer, we said it had bugs in it.

*W*hen differential geometer Kang-Tae Kim was still a student, he met the great Chinese geometer S. S. Chern. Chern had many accomplishments to his credit, including seminal contributions to the differential geometry of several complex variables.

S. S. Chern

Now Chern was a really nice guy, and he loved students. In an effort to be friendly, Chern said, "And how did you do in your complex analysis course?" Kim was embarrassed, because this was the one math course in which he had slipped up. He said, "My grade was C." Chern smiled and said, "I beat you. I got a C+."

*W*illiam Paulsen (1963–) was a graduate student in mathematics at Washington University in St. Louis. He hailed from South Dakota, and indeed attended a small college in Sioux Falls. So Bill was most comfortable with a low-key, rather bucolic lifestyle, and was occasionally nonplussed by life in St. Louis.

One day Paulsen and some "city friends" were comparing world views and Bill was moved to say, "You know, there is very little serious crime where I come from. I will say, however, that in the Fall you have to lock your car." A moment of puzzled silence ensued. "If you don't," continued Bill, "people will leave zucchini on your back seat."

*F*or the past many years, and especially since the advent of NSF grants, we have been living under the specter of "publish or perish." The meaning of this aphorism is that, if you are an academic, and if you want to get tenured or promoted, or you want to get a grant, or you want an invitation to a conference, or you want a raise, or you want the respect and admiration of your colleagues, then you had better publish original work in recognized, refereed journals or books. Otherwise you're outta here. Who coined the phrase "publish or perish"?

One might think that it was a President of Harvard. Or perhaps a high-ranking officer at the NSF. Or some Dean at Cal Tech. One self-proclaimed expert on quotations suggested to me that it was Benjamin Franklin! But, no, it was Sociologist Logan Wilson in his 1942 book *The Academic Man: A Study in the Sociology of a Profession* [WIL]. He said, "The prevailing pragmatism forced upon the academic group is that one must write something and get it into print. Situational imperatives dictate a 'publish or perish' credo within the ranks."

Wilson was a President of the University of Texas and a student at Harvard of the distinguished Sociologist Robert K. Merton. So he no doubt knew whereof he spoke.

The estimable Marshall McLuhan has sometimes been credited with the phrase "publish or perish," and it is arguable that he popularized it. In a June 22, 1951 letter to Ezra Pound he wrote (using Pound's favorite moniker "beaneries" to refer to the universities)

The beaneries are on their knees to these gents (foundation administrators). They regard them as Santa Claus. They will do 'research on anything' that Santa Claus approves. They will think his thoughts as long as he will pay the bill for getting them before the public signed by the profesorry-rat. 'Publish or perish' is the beanery motto.

*W*hen Alfred Tarski (1902–1983) joined the faculty at U. C. Berkeley in 1942, Jerzy Neyman (1894–1981) was already well established there as Director of the Statistics Laboratory. It was expected that Tarski and Neyman—fellow refugees from Poland—would strike up a strong camaraderie, but in fact this never took place. They had different political views, but perhaps more important was the fact that they were both ambitious empire-builders. They tended to get in each other's way. It was a standard joke in Berkeley to describe Tarski and Neyman as "Poles apart."

J once co-hosted an education conference at MSRI (the Mathematical Sciences Research Institute in Berkeley). Right before introducing a particular speaker, I asked her, "What is your title?" Expecting to hear something like *New Trends in Calculus Teaching*, I instead got a shrill version of "My title is *Professor*, but you can call me ...."

*W*e all wrestle with the conundrum of what to publish. Coupled with the imperative that we *must* publish, this sometimes puts us in quite a spot.

A mathematician once told a student, "If you understand a theorem and you can prove it, publish it in a mathematics journal. If you understand it but can't prove it, submit it to a physics journal. If you can neither understand it nor prove it, send it to a journal of engineering."

Eminent physicist Freeman Dyson (1923–) has a different spin on the matter: "Most of the papers which are submitted to the *Physical Review* are rejected, not because it is impossible to understand them, but because it is possible. Those that are impossible to understand are usually published."

*H*endrik Lenstra, late of U. C. Berkeley and currently located in Leiden, is one of the great algebraists of our time. The recent movie *Porridge, Pulleys, and Pi* by George Csicsery gives an account of his life and development.

Lenstra has many interests, ranging from the art of M. C. Escher to rigorous swimming for his health. Speaking of his style of doing mathematics, Lenstra says, "I don't like to calculate. It interferes with my lap count."

$\mathcal{C}$ertainly the driving force behind the function theory of several complex variables in the first half of the twentieth century was the solution of the Levi problem—to give an extrinsic geometric characterization of domains of holomorphy. The problem was finally cracked by Kiyoshi Oka (1901–1978) using the theory of the Cousin problems.

Well, the French got ahold of this stuff and abstracted the heck out of it. Jean Leray (1906–1998), Henri Cartan (1904–), Jean-Pierre Serre (1926–) and others created the theory of coherent analytic sheaves which nicely formalizes many of Oka's idea. Further, coherent analytic sheaves are a seminal tool in their own right.

One day the powerful German complex analyst Hans Grauert (1930–) asked Oka what he thought of coherent analytic sheaves. The reply: "Darkness has finally descended upon mathematics."

$\mathcal{T}$he first volume of *Mathematical Apocrypha* told of the exploits of Frank Wilczek (1951–). Frank used to be at the Institute for Advanced Study but recently moved to MIT. His latest triumph is that he won the Nobel Prize in Physics (2004).

As we related before, Frank lived in Albert Einstein's house on Mercer Street in Princeton. Algebraic topologist Michael Davis (1949–) went, with his two children, to visit Frank. They spent the afternoon in pleasant repartee, and then returned home for dinner. That evening, the two kids started talking animatedly of Beavis and Butthead. One kid would describe some scatalogical activity performed by one of the characters, and then the other kid would top it. Mike interrupted, asking what in the world they were discussing. "It's *Beavis and Butthead*, Dad!" Mike Davis was perplexed. "What's that?" The kids indulged in the usual frustration with slow parents and said, "It's the new show on MTV." Now Mike was even more confused. "But we don't have cable. How could you have seen this show?" In chorus the kids shouted back, "We saw it at Einstein's house!"

$\mathcal{M}$athematics is a special case of mathematical economics.
—Gérard Debreu, Nobel Laureate in Economics, 1983

$\mathcal{L}$ouis Nirenberg (1925–) has a concept of "mathmanship." This is where someone asks a pointed question after your colloquium talk and you trump the questioner with an even more pointed answer. As an instance:

Interrogator: Didn't that result appear in Gauss?

Speaker: Yes, but it was wrong.

$\mathcal{T}$he State of California has had a long sequence of controversial and out-spoken Superindents of Public Instruction. Bill Honig (1937–) was one of these. He once said that he thought the general public should have a voice in defining what an excellent teacher should know. "I would not leave the definition of math up to the mathematicians."

$\mathcal{T}$he estimable Saunders Mac Lane was so taken with Andrew Wiles's (1953–) proof of Fermat's last theorem that he wrote a poem for the occasion. That literary work is in fact so lengthy as to preclude inclusion in this short, little book. But we can quote a few stanzas of the work "*Will Fermat Last?*":

> Once Diophantus
> Lent us his name
> This action gave him
> Posthumous fame
>
> Add up two nth powers
> Get back just one
> Using whole numbers
> Would make lots of fun
>
> Pythagoras knew this
> When n is just two
> But for higher exponent
> There was no proponent

Then Fermat succeeded
With exponent four
Continued to study
To get something more

…

If abroad,
A. Grothendieck
Is just the man
With whom to speak

Artin, Mazur,
Other guys
Travelled there
And got quite wise

Now it's rather
Pierre Deligne
Of whose talents
We should sing

For the Weil conjecture
Yielded
To techniques
That he fielded

…

While the ghost
Of Fermat smiles
Give three cheers
For Andrew Wiles

So many parts
A hole in one
We trust repair
Can soon be done

Tom Apostol (1923–) was also moved to write a poem in homage to Andrew Wiles. His opus, "*Ode to Andrew Wiles, KBE*," is brief enough to include *in toto*:

> Fermat's famous scribble—as marginal note—
> Launched thousands of efforts—too many to quote.

Anyone armed with a few facts mathematical
Can settle the problem when it's only quadratical.
Pythagoras gets credit as first to produce
The theorems on the square of the hypotenuse.

Euler's attempts to take care of the cubics
Might have had more success if devoted to Rubik's
Sophie Germain then entered the race
With a handful of primes that were in the first case.

Lamé at mid-century proudly announced
That the Fermat problem was finally trounced.
But the very same year a letter from Kummer
Revealed the attempt by Lamé was a bummer.

Regular primes and Kummer's ideals
Brought new momentum to fast-spinning wheels.
Huge prizes were offered, and many shed tears
When a thousand false proofs appeared in four years.
Then high-speed computers tried more and more samples,
But no one could find any counter examples.

In June '93 Andrew Wiles laid claim
To a proof that would bring him fortune and fame.
But, alas, it was flawed—he seemed to be stuck—
When new inspiration suddenly struck.

The flaw was removed with a change in approach,
And now his new proof is beyond all reproach.
The Queen of England has dubbed him a Knight
For being the first to show Fermat was right.

*Y*et another paean to the solution of Fermat's problem was penned by E. Howe, H. Lenstra, and D. Moulton. It goes like this:

"My butter, garçon, is writ large in!"
A diner was heard to be chargin',
"I had to write there,"
Exclaimed waiter Pierre,
"I couldn't find room in the margarine."

Sherman Stein of the University of California at Davis is one of the more successful calculus authors of our time.

These days, the writing of a calculus book is something of an ordeal. There are myriads of reviewers, and the author must jump through scores of hoops to please editors, reviewers, marketing groups, and many others. Stein has a somewhat philosophical view of the matter. He is quoted as saying

> When you write a calculus book, they give you a good deal of choice. You can call it *Calculus* **and** *Analytic Geometry* or you can call it *Calculus* **with** *Analytic Geometry*.

When I was a graduate student, I was a regular fan of the CBS evening news broadcast. One of the featured journalists was Eric Sevareid (1912–1992), a rather dour and serious reporter of the old school. Each evening he would deliver a studied and sometimes hardbitten opinion piece about current events.

On one particular evening he was commenting about the Watergate scandal, and he drew a certain parallel with selected historical events. Then he said, "But parallel lines never meet, even at infinity." Ahem. I felt that I knew something about this matter, and I felt that Sevareid had erred. Next day I got some official Princeton stationery from the departmental supply cabinet; I wrote to Sevareid, pointing out that in any theory of geometry in which there *was* a point at infinity, it was generally the case that parallel lines met there. End of discussion.

To my surprise and delight, I received a reply to my letter the next week. Here is what Eric Sevareid said:

> What I actually said was, "Parallel lines don't meet, if we remember grade school geometry, even in infinity." Not: "as we all remember …"
>
> The line was a kind of "literary conceit", which you, as a mathematician, would probably not appreciate. The reason for "if" and the "grade school" was to downgrade the proposition a bit, since even I am vaguely aware of more sophisticated mathematical theories to the contrary.
>
> I notice the sun rose on schedule the next morning in spite of my cavalier behavior.

Blaise Pascal (1623–1662) once observed, "Had Cleopatra's nose been shorter, the whole face of the world would have changed."

Teaching is an important pursuit, and has its own rewards. So is research. But the two can work symbiotically together, and the whole created thereby is often greater than the sum of its parts. J. J. Sylvester (1814–1897), who was an eccentric teacher at best, describes the process in this way:

> But for the persistence of a student of this university in urging upon me his desire to study with me the modern algebra I should never have been led into this investigation; and the new facts and principles which I have discovered in regard to it (important facts, I believe), would, so far as I am concerned, have remained still hidden in the womb of time. In vain I represented to this inquisitive student that he would do better to take up some other subject lying less off the beaten track of study, such as the higher parts of the calculus or elliptic functions, or the theory of substitutions, or I wot [know] not what besides. He stuck with perfect respectfulness, but with invincible pertinacity, to his point. He would have the new algebra (Heaven knows where he had heard about it, for it is almost unknown in this continent [America]), that or nothing. I was obliged to yield, and what was the consequence? In trying to throw light upon an obscure explanation in our text-book, my brain took fire, I plunged with re-quickened zeal into a subject which I had for years abandoned, and found food for thoughts which have engaged my attention for a considerable time past, and will probably occupy all my powers of contemplation advantageously for several months to come.

I was once trying to assist one of my undergraduates at UCLA with some calculations he was doing in a part-time job he had at Hughes Aircraft. At some point it became clear that he needed to show me some of the documents that he had from Hughes; and he was rather nervous about doing so. Of course I am accustomed to working in an academic environment where ideas are generally exchanged freely and openly. I didn't know how to react. But I knew that I had entered some strange new Twilight Zone when he finally brought me a sheaf of papers with a front page that read as follows:

WARNING!

The following document has been coated with a xerographically sensitive substance. Any attempt to reproduce this document will result in the sensitization of the substance.

*R*odney Vaughan, of my hometown Redwood City, California, sums up his views of mathematics and religion:

> The one true religion is mathematics. We know this, because we observe that gods exist only in the minds of their believers; thus, their relative powers can be accurately determined simply by counting the heads of the believers. What this means is that theology is just a subset of arithmetic.

*J*ohn Horton Conway (1937–) is one of the more remarkable mathematicians of our time. The John von Neumann Professor in Applied and Computational Mathematics at Princeton, Conway is certainly one of the illuminati of Fine Hall. He spends most of every day sprawled on a sofa in the Commons Room, talking about mathematics to anyone who will stand still for a few minutes. Several years ago he gave up his office because the Math Department needed the space, and he did not.

John Horton Conway

Conway always has some tricks in his pocket, or on the table, to show people. On one recent visit he corralled me, took me to another room, and began setting up pennies on edge on the table. He quickly demonstrated that, if you pound your hand on the table, then most of the pennies come up tails. But if you first set the pennies spinning rapidly, then most come up heads.

All of Conway's tricks and demonstrations are combinatorial, or algebraic, or sometimes physical. They all have analytical explanations, and John Conway wants to understand everything.

He is a captivating and energetic teacher. He has been observed clawing the air and roaring like a lion to get the attention of his class. Sometimes, in preparation for presenting a new theorem, he will fling himself at full length on a table, close his eyes, and then leap up and exclaim, "This theorem is so wonderful I just needed to rest before I tell you about it. It is stupendous! When I taught it last year, I jumped up and down so hard I split my trousers!" Conway begins class by entering the room and throwing off his sandals; he signals the end of class by putting them back on.

Several years ago, Conway and Bill Thurston (1946–) ran a course called "Geometry and the Imagination". The course involved many props, geometric models, and a certain amount of tomfoolery. The course had more than 200 students—many of them people from the town of Princeton. It was a participatory event, and a great many people came away educated and enlightened.

At math conferences, Conway is a stirring lecturer. Many of his (serious) math talks involve strenuous audience participation, and there is always a marvelous climax in which some knot is unknotted, or some braid unwound, or some other surprising geometric construct revealed.

John Horton Conway is of course the creator of the Game of Life, and he has many important mathematical creations—from diverse fields—to his credit. He is a one-man Pied Piper for mathematics, and a role model for us all.

The last theorem in John Horton Conway's book *On Numbers and Games* is

Theorem 100: This is the last theorem in this book.

The proof is obvious.

$\mathcal{J}$ohn Horton Conway's office—when he had an office—was incredibly chaotic. There were papers and books scattered everywhere. Puzzles, games, charts, novelties, and other paraphernalia were piled and heaped and stacked on all horizontal surfaces. Conway realized he had a problem when he could no longer lay his hands on his latest theorem, or list of problems, or new conjecture. He set about to design a physical device that would address his quandary, and imbue some order into the chaos. After some time and effort, he had produced a set of plans. He was about to go off and find a craftsman to implement his idea in wood and metal when he noticed that such an item was already standing, empty and unused, in the corner of the office. It was the filing cabinet!

$\mathcal{T}$he Math Department at Washington University has a scientific computer lab for faculty and graduate students. Prominently exhibited on the door is a newspaper headline that says

>   Indiana State Police cracking down on math lab operations

One might surmise that there is a typographical error in the seventh word, second position.

$\mathcal{L}$ance Small (1941–) of U. C. San Diego was once teaching a probability course. As an object lesson, corollary to the law of large numbers, he told his students that it is idiotic to play the lottery. One of the students in class had parents who had just won $7 million in the Colorado State Lottery. She leaned over to her friend and said, "Should I tell him?" The friend said, "Better tell him after the final exam." So they waited until the next semester and marched to his office and let him know that at least *some* people had done quite well with the lottery. He said later—to a friend—"I'm still right."

$\mathcal{J}$n the early part of the twentieth century, the MIT Math Department had several faculty who specialized in teaching. One of the most distinguished of these was Leonard Macgruder Passano (1866–1943). He was a Baltimore man, and author of a number of important textbooks. He had a neatly trimmed beard and dressed immaculately (he even wore spats!). He had a large reproduction of Manet's *Olympia* above the desk in his office. He had strong opinions. On one occasion, when plans were discussed to strengthen the applied side of the department, he opposed it (in a memo) by saying, "Mathematics, the queen of the sciences, should not become its quean."

*I*n 1955, as performance at a party for a number theory conference held at Cal Tech, Tom Apostol wrote a poem/ditty about the Riemann zeta function. Here we include just some of the verses (it is to the tune of *Sweet Betsy from Pike*):

Where are the zeros of zeta of *s*?
G. F. B. Riemann has made a good guess,
They're all on the critical line, said he,
And their density's one over $2\pi \log t$.

This statement of Riemann has been like a trigger,
And many good men, with vim and with vigor
Have attempted to find, with mathematical rigor
What happens to zeta as mod *t* gets bigger.

The names of Landau and Bohr and Cramér,
And Hardy and Littlewood and Titchmarsh are there,
In spite of their efforts and skill and finesse,
In locating the zeros no one's had success.

In 1914 G. H. Hardy did find
An infinite number that lay on the line,
His theorem, however, won't rule out the case
That there might be a zero at some other place.

Oh, where are the zeros of zeta of *s*?
We must know exactly. It won't do to guess.
In order to strengthen the prime-number theorem,
The path of integration must not get too near 'em.

There are other verses, and some have been added. One particularly charming couplet is

Related to this is another enigma
Concerning the Lindelöf function $\mu(\sigma)$.

Jn the 1940s, Paul Smith (1900–1980) of Columbia University was a dedicated topologist, quiet and concentrated by nature. He is rumored to have said, "Whenever I see a derivative it gives me nausea." He was instrumental in bringing Sammy Eilenberg (1913–1998) to Columbia.

Certainly one of the more influential analysis books of modern times has been *Introduction to Fourier Analysis on Euclidean Spaces* by E. M. Stein and Guido Weiss. One of the many math departments that have fallen under the spell of that book is the University Autonoma de Madrid. The department is peopled by students and collaborators and acolytes of Stein and Weiss; it is a hotbed of analysis.

If one is a guest of the department, and stays at a hotel in Madrid, then one takes the train out to the periphery of the city where the university is located. The name of the metro stop at which one exits the train is "Capobianco"—which means (transliterating from Latinisms to Germanisms) Stein-Weiss.

Jn the book *Mathematics: People, Problems, Results* [CAH], the authors introduce S. L. Segal's (1937–) article on Helmut Hasse (1898–1979) with these words:

> The mathematician as superman is a common theme in biographical works by mathematicians about mathematicians. Always wiser, more truly moral, kinder, more broad-minded, less petty, transported above the niggling concerns of mortals on the wings of pure reason, the mathematician is both hero and savant. Anyone who wants to really understand the essential nature of almost anything should ask a mathematician. No mathematician ever fell for a crude political scheme, or rallied behind a demagogic rabble-rouser, or unwittingly bought swampland in Florida. The mind that grasps, generalizes, and extends calculus as an evening's entertainment can surely discern the nonsense and sham so readily accepted by the common folk. That is too often the way mathematicians prefer to view themselves. Reality, however, tends to be rather less charitable.
>
> Consider, as exhibit A, the eighteenth-century French geometer who liquidated a substantial portion of his considerable fortune to purchase antiquities such as an original letter, in French no less, from

the Apostle John to the Virgin Mary. Or exhibit *B*, the contemporary mathematician who, when convinced of the imminence of atomic warfare, spent a whole week in his backyard with pick and shovel constructing an earth-covered fallout shelter. The structure was well reinforced with timbers and covered with three feet of earth and lacked nothing but a way to get inside it. There is a virtually endless supply of such eccentric anecdotes, but why bother? No one but mathematicians believe the first description anyway. Beyond such eccentricity lies the rather gritty and much less charming reality that mathematicians are no more noble or virtuous than the rest of us.

A few years ago Landon Clay (descendent of the Civil War general Cassius Clay), endowed the Clay Mathematics Institute. The Clay Institute is loosely affiliated with the Harvard University Mathematics Department. Its $50 million endowment gives it a considerable income each year. And one of the many outstanding things that they have done with this money is to pose the seven Millennium Prize Problems. These are seven mathematics problems, each with a $1 million bounty attached. Sketchily described, the problems are these:

1. The Birch and Swinnerton-Dyer Conjecture [algebraic geometry]
2. The Hodge Conjecture [complex geometry]
3. The Navier-Stokes Equations [mathematical physics, partial differential equations]
4. The **P vs. NP** Problem [logic, theoretical computer science]
5. The Poincaré Conjecture [topology]
6. The Riemann Hypothesis [analytic number theory, complex analysis]
7. The Yang-Mills Theory [mathematical physics, partial differential equations]

The Clay Institute has detailed and strict rules for how purported solutions to these problems will be judged. A portion of the strictures is thus:

Before consideration, a proposed solution must be published in a refereed mathematics journal of world-wide repute, and it must also have general acceptance in the mathematics community *two years* [after that publication]. Following this two-year waiting period, the SAB [Scientific Advisory Board of the Clay Institute] will decide whether a solution merits detailed consideration … The SAB will pay special attention to the question of whether a prize solution depends crucially on insights published prior to the solution under consideration. The SAB may (but need not) recommend recognition of such prior work in the prize citation, and it may (but need not) recommend the inclusion of the author of prior work in the award.

Thus far there is only one serious contender for a Clay Prize, and that is Grisha Perelman's solution of the Poincaré Conjecture.

*O*f course mathematics has more than its fair share of controversy. In recent years, considerable vituperation has arisen from various alleged proofs of the Kepler sphere-packing conjecture. Recall that the problem is to determine the most efficient method for packing spheres of the same size in 3-dimensional space. [The analogous 2-dimensional problem—for packing discs of the same size in the plane—is known to have the standard hexagonal packing as its optimal solution.] In 1993, Wu-Yi Hsiang (1937–) of the University of California at Berkeley published a proof of the 3-dimensional conjecture.[*] There had been no significant work on the problem for a long time, so Hsiang's paper took the world by storm. Unfortunately, there was not general agreement on the completeness or correctness of Hsiang's arguments. It seems that he only considered certain special cases, which he claimed were the "worst-case scenarios." People were not convinced.

One of the people who led the attack on Hsiang's work was Thomas Hales (1958–) of the University of Michigan. In his 1994 article in the *Mathematical Intelligencer*,[†] Hales went into some detail to discredit Hsiang's efforts. Other notable mathematicians, some at Princeton University, were fairly vocal in critizing Hsiang's work.

It turns out that Hales was laying the foundation for his own work on the problem. Soon thereafter, Tom Hales published a series of papers—between 1992 and 2002—which (it was advertised) were stepping-stones to the solution of the Kepler problem. Finally, in 2002, Hales wrote the definitive paper proclaiming a solution—a 200-page manuscript with arguments that rely on hours and hours of computer calculation. He submitted this work to the venerable *Annals of Mathematics*.

---

[*] See W.-Y. Hsiang, On the sphere packing problem and the proof of Kepler's conjecture, *Internat. J. Math.* 4(1993), 739–831.

[†] See T. Hales, The status of the Kepler conjecture, *Math. Intelligencer* 16(1994), 47–58.

Now it is well known that the *Annals* does not publish very long papers—ones that would take up a whole issue by themselves. And it tends to concentrate on mainstream mathematics—not 350-year-old stuff that nobody thinks about anymore. But the *Annals* embraced Hales's paper. Editor Robert MacPherson sent the paper to a twelve-man team of Hungarian mathematicians headed by L. Fejes-Toth (son of the Fejes-Toth who pioneered work on Kepler's problem), asking them to referee the voluminous work.

Well, the Hungarians spent a full year running a seminar and trying to battle their way through Hales's behemoth paper. But they found that it was so heavily reliant on computer calculation that they could not, in the end, certify its correctness. And so they told MacPherson.

But MacPherson seems to be quite enamored of computer proofs. He decided to nonetheless accept the Hales paper for the *Annals*, and to publish it with a disclaimer saying that the journal could not be sure that the result was correct.

When I heard this story I was nonplussed (to say the least). I contacted John Horton Conway at Princeton, a well-known expert on Kepler's sphere-packing problem and obviously someone who is very close to the fire. He said that he had been having open-heart surgery and had not heard about any of these developments. But he, too, was perturbed by the news. He contacted MacPherson and convinced him that a disclaimer was highly inappropriate for a much-touted paper to appear in the *Annals*. So now the *Annals* has replaced the disclaimer with a ringing endorsement.

It is good to know that one has had a small influence on the history of modern mathematics.

*J*ohannes Kepler (1571–1630) is best remembered for his three laws of planetary motion, which he derived with bare-hands calculations from the masses of data left behind by his teacher Tycho Brahe (1546–1601). Kepler's life was an unrelenting struggle filled with hardship and sorrow. His wives died young from the deprivations of poverty, and Kepler himself always suffered from ill health. In his fifty-eighth year he had a premonition of his own death and he wrote this epitaph for himself:

> I used to measure the heavens,
> now I measure the shadows of the Earth.
>
> Although my mind was heaven-bound,
> the shadow of my body lies here.

Kepler died a few months later. He was buried in the graveyard outside the church at St. Peters, Regensburg, Germany. But, a few years after his death, the Thirty Years War ravaged the area and Kepler's burial site was permanently obliterated.

$\mathcal{H}$arold Bacon (1907–1992) remembered the great Edmund Landau in this way:

> Landau was a man of commanding presence with a real sense of humor, an enthusiastic lecturer, meticulously dressed in a somewhat formal fashion. He was particularly annoyed by chalk dust. In those days the blackboards in our department classrooms [at Stanford University] were of black slate, and we had rather soft chalk—white, yellow, red, green and other colors. Landau would write in unusually large script, quickly filling the front blackboards. He would sometimes dart about the room and write on the sidewall blackboard— once he even climbed over a couple of chairs to get to the board on the back wall. But then, the boards [had to] be erased so that the writing could go on. Landau abhorred the usual felt erasers—too much dust. So, on the first day of his 8:00 and 9:00 classes, his assistant brought in a granite-ware kettle in which were a sponge and some water. Since she adamantly refused to use the sponge on the blackboard, Landau himself (shades of Göttingen with assistants who did the erasing!) would grasp the sponge, wring the water out on the floor, make some passes at the board, and then call on one or two students or visitors to his lecture to come up and dry off the slate with paper towels. A very ineffective method of drying! The lecture would continue. But the slate was still slightly wet, so half the chalk marks didn't show. Eventually the board dried, however, and normal conditions returned. But the paper towels usually got on the floor where they mingled with water and various scraps of white, yellow and red chalk. After two hours of being walked on, these additions to the bare wood floors produced a, shall we say, cluttered appearance. As it happened, the 10:00 class that followed these first two classes in this classroom was a course in Education—something like "The administration and care of the School Building and Classrooms." I fear that those students had a rather spectacular illustration of the neat and orderly classroom! Incidentally, on the last day of classes, Landau

made a graceful and humorous farewell speech in his heavily accented English. His last "goodbye" ended with the request, "Please preserve the sponge to remember me by."

*H*arold Bacon also recalls warmly his relationship with Dunham Jackson (1888–1946). Evidently Jackson was an "inspired composer of Limericks of the quite respectable sort." Bacon purchased Jackson's book *The Theory of Approximations* and took it into Jackson's office in order to request an autograph. In fact Bacon suggested that Jackson write a Limerick. Jackson immediately picked up the book and, without any cogitation, wrote on the flyleaf:

> There was a young Fellow named Bacon
> Whose judgment of books was mistaken
> In a moment too rash
> He relinquished some cash
> And his faith in the Author was shaken
> (August 17, 1934)

Bacon adds that his faith in the author was by no means shaken; it was greatly reinforced.

*H*alsey Royden (1928–1993) was characteristically modest when recalling his own exploits. He said, "It had always been my intention to go East for doctoral work after finishing my Masters Degree at Stanford. Don Spencer (1912–2001) and Paul Garabedian (1927–) both told me to go to Harvard and work with Lars Ahlfors (1907–1996). Although that may not have been the best advice I have ever been given, it was certainly the best advice I ever accepted. Paul also instructed me to introduce myself to Stefan Bergman (1895–1977) and to ask him for a research assistantship. This I did upon my arrival at Harvard, and my acquaintance with Garabedian, Schiffer (1911–1997), and Spencer was sufficient to obtain an appointment as one of Stefan's assistants."

*S*aunders Mac Lane speaks fondly of the American Mathematical Society in the early days (the 1930s). Membership was relatively small, and the meetings intimate. The talks were attended by *everyone*, and interaction was vigorous. Mac Lane's experiences with the AMS were not always salubrious, however. He tells of one particular incident at the AMS meeting in Cambridge, Massachusetts in 1933:

> I needed a job for the next year, so I announced and gave a 10-minute paper on logic, entitled "Abbreviated proofs in logic calculus". As soon as my 10 minutes were over and the chairman had asked for questions, Øystein Øre (1899–1968) rose and spent the next 10 minutes denouncing my work. Mathematical logic (and even more, philosophical considerations) did not in his view belong in meetings of the AMS, and he made this point very clearly. It was not really at my expense, since George D. Birkhoff (1884–1944) and other Harvard professors were in the audience, and voted a few months later to offer me for the following year an appointment as Benjamin Peirce Instructor at Harvard (I accepted with alacrity). The paper on logic which I had then presented was later published ... and soon forgotten; it was not profound and may well have deserved Øre's criticism."

To those of us who are ordinary mathematicians, there appear to be certain towering figures in the discipline who lead the way and establish all the major results. They appear to be God's anointed ones, and they play a special and seminal role in the profession. Evidence is that there is more consensus among mathematicians as to who the important workers are than there is in almost any other discipline. Thus we are perhaps more conscious of a pecking order than are literary critics or historians. Certainly John Milnor (1931– , Fields Medal, 1962) is one of the extraordinary mathematicians of our time. Here is his view of the mathematical food chain:

> What I love most about the study of mathematics is its anarchy! There is no mathematical czar who tells us which direction we must work in, what we must be doing. There are thousands of mathematicians all over the world each going in his or her own direction. Many are exploring the most popular or fashionable directions, but others work in strange or unfashionable directions. Perhaps many are going the wrong way, but cumulatively the many different directions, the many different approaches, mean that new and often unexpected things will be discovered. I like to picture the frontier of mathematics as a great ragged wall, with the unknown, the unsolved problems, to one side, and with thousands of mathematicians on the other side, each trying to nibble away at different parts of the problem using different

approaches. Perhaps most of them don't get very far, but every now and then one of them breaks through and opens a new area of understanding. Then perhaps another one makes another breakthrough and opens another new area. Sometimes these breakthroughs come together, so that we have different parts of mathematics merging, giving us wide new perspectives. Often the people who make these breakthroughs are those who are well known, those we expect to obtain good results; but not always. Many times major results are obtained by those who are not at all well known, or by people we may know but underestimate, so that we are completely surprised to find that they have accomplished so much. It is wonderful that no one has the power to turn such people off!

$\mathcal{H}$alsey Royden recalled the changing character of the Stanford Department Seminar. He tells it this way:

Fellow students told me that in earlier years it had sometimes been a harrowing experience for them. According to Albert Novikoff, one of our more flamboyant graduate students and now a professor at NYU, the first hour consisted of the student's attempt to present the assigned topic, while Pòlya (1887–1985) and James V. Uspensky (1883–1947) argued with each other about the adequacy of the student's statements. For the second hour Pòlya or Uspensky would demonstrate how the lecture should have been given. The only mitigation for the unfortunate student was that, whichever of Pòlya or Uspensky was critical, the other would be supportive.

During the intermission Pòlya and Uspensky would

George Pòlya

bait Schaeffer and Spencer about these young men's ignorance of classical mathematics. Schaeffer was oblivious to this, but Spencer would sometimes rise to the bait and respond by asking Uspensky what the Betti numbers of the sphere were. Uspensky would indicate that this modern stuff was nonsense beneath his notice. Sometimes the argument descended to the personal level, with Uspensky maintaining that the younger generation (i.e., Schaeffer and Spencer) lacked the strength of character and fortitude exhibited by their elders. Spencer recalls only once that he or Schaeffer got the better of the exchange: Uspensky had been holding forth about the degeneration of the younger mathematicians and recounted the story of an ancient Roman who, becoming tired of the world, ordered his servants to construct a huge funeral pyre which he proceeded to walk into. Turning to Schaeffer, Uspensky said, "Would you do that, Schaeffer?" Schaeffer bowed and replied, "After you, Uspensky."

"Logic is in the eye of the logician."

—Gloria Steinem

*D*uring a SIAM (Society for Industrial and Applied Mathematics) meeting in Ames, Iowa, in the 1960s, a group of numerical analysts (including a significant subset from England) went to a local nightclub and were much taken with a stripper named "Diane Midnight". In fact they were so enthusiastic about her performance that they rewarded her with a standing serenade of *God Save the Queen*. Meanwhile their leader, Leslie Fox (1918–1992), put his hands on his head in despair—he could be identified because he had a program of the meeting in his hand.

*D*uring the International Congress of Mathematicians in 1966 (which took place in Moscow), numerical analyst Jim Douglas (1927–) treated the Russian waiter in one of the mainline Moscow restaurants to a little touch of Texas. To wit, he looked at the plate of food that the waiter brought to the table and set before him, eyed it warily, then looked up at the man and asked in quasi-stentorian tones, "What kinda animal was this here tore offa?"

"Mathematics has given economics rigor, but alas, also mortis."

—Robert Heilbroner

$\mathcal{D}$on Spencer was one of the luminaries of post-World War II Princeton. Here is how he recalls communal life in the Princeton Math Department:

> Fine Hall was a remarkable community whose members attended a tea starting at 4:00 in the Common[s] Room. All of us knew and talked to each other and the atmosphere was usually warm and friendly. The graduate students would sometimes talk about the faculty, characterizing them in succinct and humorous ways … Lefschetz never stated a false theorem or gave a correct proof; Bochner says one thing, writes another and means a third; Church (1903–1995) is so cautious that, when he signs his name, he counts the letters in it … Bochner and Lefschetz did not get along with each other personally but each admired the other as a mathematician. Bochner said that Lefschetz was the greatest living homologist and Lefschetz said "damn it, Bochner is a fine mathematician.

$\mathcal{M}$y Ph.D. student Daowei Ma (1954–) was always anxious to improve his English, and to learn as much as he could about American culture. One day he brought me a computer disc for a text editor that would not function properly. This was in the early days of PCs, and the disc was supposed to be self-booting. I fancied myself something of an expert so I popped the disc into the machine to see what I could do.

It turned out that the `autoexec.bat` file required some adjustment. I made the necessary changes, rebooted the machine with the diskette in the `A:` drive, and it worked. I shouted, "Bingo!" Daowei looked puzzled. He said, "Bingo? What's Bingo?"So I told him about the game of Bingo, about the old ladies who play it, how the person on the stage would draw numbers and call them out, and you mark the numbers on your card. When you get five in a row you shout, "Bingo!" and you win. Ma listened very carefully and nodded seriously at the end of this explanation. "So," he said, "when you are trying to do something and it finally works you shout 'Bingo!'." I nodded. "And," he went on,"when it doesn't work you shout 'Shi—'."

*F*ields Medalist Charlie Fefferman (1949–)
tells of being at a cocktail party and chatting
with a businessman. The man of the world
asked the professor, "And how much do you
teach?" "Nine hours." was the honest reply.
"Well, that's a long day," allowed the businessman. "But it's easy work."

"All truths are half-truths."
   —Alfred North Whitehead (1861–1947)

Charles L. Fefferman

A friend of mine, Harold Parks (1949–), was studying for his French language exam as part of the PhD program at Princeton. Parks did not know
any French at all, so he decided to learn mathematical French by reading a
French math book. He selected a particular tome that happened to begin
every proof and every discussion with the phrase "Nous sommes ...." He
also had the world's worst French/English dictionary. He came to me one
day and queried, "Why does every proof in this book begin with the phrase
'We beasts of burden...'?"

*L*eslie Kay (1951–) is a mathematician at Virginia Polytechnic Insitute in
Blacksburg, Virginia. Leslie has always had a flair for languages; when she
was an undergraduate she took a lot of French.

   On her first trip to Paris, Leslie was confident of her ability to communicate with the natives. One day she approached a Parisian on the street to
get directions. She formulated her query in rapid and fluent French. The
Parisian drew himself up haughtily and declared (in English), "I don't speak
Flemish."

"Only two things are infinite, the universe and human stupidity, and I'm not
sure about the former."

                                              —Albert Einstein (1879–1955)

*A*s calculus teachers, we are all put in the position of telling our students what work is and how it is calculated using the integral. Here is Bertrand Russell's (1872–1970) take on the matter: "Work is of two kinds: first, altering the position of matter at or near the earth's surface relative to other matter; second, telling other people to do so."

*W*hen Bertrand Russell had, by his second wife, a first child, a friend accosted him with, "Congratulations, Bertie! Is it a girl or a boy?" Russell replied, "Yes, of course, what else could it be?"

*T*he publication of *Principia Mathematica* [WHR] by Alfred North Whitehead and Bertrand Russell was a major event both of a literary and a philosophical, and of course also a mathematical, nature. The world did not soon tire of offering commentary, criticism, and praise of the monumental work. The topologist J. H. C. Whitehead was often asked for his views on the work of his uncle A. N. Eventually he developed a stock answer. When asked, "What do you think of your uncle's philosophy?", he would reply, "I really haven't thought much about it—but what do you think of *your* uncle's philosophy?"

*T*he book *Principia Mathematica* is a masterpiece of mathematical abstraction. Its goal was to put mathematics on a rigorous abstract footing. Wags liked to point out that the *Principia*'s proof that $2 + 2 = 4$ was impenetrable to most laymen, and to many mathematicians as well. Russell himself poked some fun at the whole business with the following tomfoolery:

> You are quite right, except in marginal cases—and it is only in marginal cases that you are doubtful whether a certain animal is a dog or a certain length is less than a meter. Two must be two of something, and the proposition "2 and 2 are 4" is useless unless it can be applied. Two dogs and two dogs are certainly four dogs, but cases arise in which you are doubtful whether two of them are dogs. "Well, at any rate there are four animals," you may say. But there are microorganisms concerning which it is doubtful whether they are animals or plants. "Well, then living organisms," you say. But there are things of

which it is doubtful whether they are living organisms or not. You may be driven into saying, "Two entities and two entities are four entities." When you have told me what you mean by "entity," we will resume the argument."

Utter Sagacity

*I*n our modern world, we are somewhat spoiled by our many conveniences and amenities. I was once at dinner with Walter Rudin (1921–) and he told me how the writing of his celebrated book *Principles of Mathematical Analysis* came about. Namely, he was a Moore Instructor at MIT, had to teach undergraduate real analysis, and lamented that there was no proper textbook. Somebody said, "Walter, you ought to write one." And the rest is history.

I asked Walter what people used for texts prior to the 1950s, and his somewhat startling reply was, "There were sets of notes circulating around." Of course I believed Walter, but I still found his answer incredible. Walter's assertion is bolstered by this passage from Halmos:

> The dissemination of mathematical information in the 1930s and 1940s was not as efficient as it has since become. Xerox copies and preprints had not been invented yet, the number of new books that appeared each year was finite, and the everywhere dense set of meetings and conferences that we live with today was still far in the future. One ingenious system that did exist was that of "notes." If a good mathematician gave a good course at a good university, the notes for the course were in great demand, and, before long, the system of producing and distributing such notes became a matter of routine. A student or two would take notes during the lecture, the result might or might not be examined, changed, and approved by the lecturer (usually it was), and then it would be mimeographed (how many students in the 1980s have that word in their vocabulary?), and sold. There was no advertising except word of mouth; the usual price was $1.00 or $2.00 or $3.00. The package that reached your mailbox consisted of, say, 200 pages of what looked like typewritten material. One of

the earliest, most famous, and best examples is the ... Bohnenblust notes (in real analysis, 1942–1943).

Certainly one of the treasures of the Fine Hall Mathematics Library in Princeton is the magnificent collection of old sets of notes, as described in the preceding passage from Halmos.

"Good teaching is one-fourth preparation and three-fourths theatre."

—Gail Godwin

*I*n the first volume of *Mathematical Apocrypha* we related the tale of Henri Poincaré having discovered special relativity about the same time as— indeed slightly earlier than—Albert Einstein (1879–1955). Perhaps even more surprising was that there was a major campaign, conducted over a period of at least five years, to obtain the Nobel Prize in Physics for Poincaré. In fact Gösta Mittag-Leffler spearheaded the effort, and he had support from Paul Painlevé (1863–1933), Gaston Darboux (1842–1917), and Ivar Fredholm (1866–1927). Here is a portion of Mittag-Leffler's arguments, contained in a letter to one Professor Paul Appell (1855–1930):

> The time is come when we can hope to make Poincaré winner of the Nobel Prize. I send enclosed with the next mail a proposal written by Fredholm that he subjects to your judgment and one by Mr. Darboux. He has made considerable use of the proposal made by Darboux this year. The most important thing is first to establish the prominent part played by pure theory in physics and then to conclude with the proposition to give the prize for discoveries defined by a sufficiently simple formula. After some discussion, we have found this formula in Poincaré's discoveries concerning the differential equations of mathematical physics. I think that we will win with this program.

Mittag-Leffler went on to say that the nominators had to avoid "mathematics" and refer to "pure theory" because "like those who are only experimentalists, members of the Nobel committee for Physics are scared silly by mathematics."

Needless to say, Poincaré was never awarded the Nobel Prize. In one year, thirty-four eminent physicists and mathematicians supported the nomination. To no avail; the experimentalists were never convinced. Mittag-Leffler summarized the situation as follows:

We have again been beaten, this time for the Nobel Prize. This crowd of naturalists who do not understand anything about the fundamentals of things has voted against us. They fear mathematics because they don't have the slightest possibility of understanding anything about it.

$\mathcal{J}$ohn M. Worrell, Jr. (1933–) was one of R. L. Moore's (1882–1974) most ardent admirers. He seemed to want to be as much like the great man as he could manage. In the movie *Challenge in the Classroom*, which recounts lovingly of R. L. Moore, his teaching method, and his students, Moore tells of a certain "Mr. W." who had the habit of always leaving the room when others were presenting their results. He didn't want to be "tipped" on how to do the problems and proofs. Worrell got an MD before earning his PhD at the University of Texas. Like many of Moore's students, and like Moore himself, Worrell expressed himself with great care. At a topology conference in Baton Rouge (Moore himself is credited with inventing the phrase "point-set topology"), Worrell was speaking of Moore. He said, "I never knew his equal." Then he paused to consider what he said. After some thought he added, "Except for Moore himself."

$\mathcal{M}$arston Morse (1892–1977) had a profound influence on twentieth-century geometric analysis. He is remembered fondly by all who knew him. Raoul Bott, as a young mathematician, had Morse as his mentor. He recalls being able to go to Morse in times of need. In 1950 he went to the great man for advice on obtaining a summer job. He was somewhat bewildered when, after considerable reflection, Morse recommended mowing lawns and baby-sitting.

When it was time, in 1951, for Bott to move from the Institute for Advanced Study to something more permanent, Morse recommended the "mathematical wasteland." More precisely, what he said to the young man was, "Get away from Princeton and the mathematical centers. Have your own thoughts in peace and quiet. Look what Lefschetz did in Kansas."

William Transue (1914–) was Morses's acolyte for many years, and reminisces about what it was like to work with Morse:

> Several years later we undertook the study of integral representation of bilinear functionals and allied topics which continued for some years. I spent one full year at the Institute and many summers with

him at various places—Princeton, Cape Cod, Maine. Working with Marston (for me at least) meant being completely taken over, spending almost all waking hours with him, talking mathematics all day, including during many meals taken with the Morse household, and continuing late into the evening. He was a real bear for work. In Princeton we usually worked in his bedroom (Richard Arens once remarked that he couldn't concentrate with someone else's pants hanging in the closet!), and in Maine sometimes in the car. Louise was always extremely good about finding him a quiet place to work, but with the number of children about, this was not always easy.

$\mathcal{A}$ famous professor at the Tata Institute in India (name withheld by request) produced a PhD student who went off to make his way in the world. After a time, the Professor received a letter from the student proclaiming that he had found a marvelous proof that every ring is a principal ideal domain (PID). "Very interesting," allowed the Professor. "But what about the ring of polynomials in two variables and the ideal generated by x and y?" The student wrote back with, "But that's just one counterexample! If you will only read my proof …"

$\mathcal{L}$eo Sario (1916–) was my colleague at UCLA for a number of years. We were pretty good friends, and I admired him both for his mathematics and for his courtesy and bonhomie. But Leo was eccentric, and this feature manifested itself in a number of ways.

Leo told me that his favorite mode of doing mathematics was on the beach at night. Around 2:00 in the morning he would take the Pacific Coast Highway to his favorite spot, and then remove from the car a fiberglass baby carrier, his papers and writing equipment, and a miner's helmet. Finding an isolated position in view of the ocean, he would seat himself, wear the baby carrier backward, put his papers in the baby carrier, mount the miner's helmet on his head, and get to work. Judging from Sario's collected work on MathSciNet, one can only conclude that this is an extremely effective method for doing mathematics.

*W*hile Sario was a UCLA Professor in the 1970s, he was a landlord. He owned some apartment buildings in Santa Monica. One year the City Government endeavored to levy a huge tax against the property holders of the town, and Sario decided to act. He screwed up all his scholarly skills and spent a couple of months doing research in the Santa Monica archives. The result was that he found that the new tax code ran afoul of an obscure and rarely cited city code that had been established fifty years before. He presented his findings at a City Council Meeting, and *he prevailed.* Leo Sario was on the front page of the *Santa Monica Evening Outlook* and he became the hero of all the other landlords in town. This was one of his proudest achievements.

*T*arski is the author of an article entitled Truth and Proof that appeared in the 1969 *Scientific American*. This piece may have the distinction of being the only article (by a mathematician) to ever appear in that period-ical intact—without editorial changes to the author's original text. The in-house editor of the piece, Joseph Wisnovsky, knew of Tarski's reputation as a stickler for details, so he did not dare tamper with Tarski's prose.

Alfred Tarski

Except for one thing: It seemed that Tarski's use of "which" vs. "that" was consistently incorrect. The editor tried to make the necessary corrections. Tarski changed them all back. The hapless editor then phoned Tarski up and read him the relevant (quite long!) article from H. W. Fowler's authoritative *A Dictionary of Modern English Usage.* Tarski listened patiently to the very end and then said, "Well, you see, *that* is Fowler. *I* am Tarski." End of dis-cussion. Tarski's article appears in the June, 1969 issue of *Scientific American* exactly as Tarski wrote it.

*E*very mathematician learns—usually as a child—about the seven bridges of Königsberg and Euler's study of them that led to the foundations of topology. Perhaps less well known—a real piece of esoteric mathematical trivia—is the names of the seven bridges: Krämer, Schmiede, Holz, Hohe, Honig, Köttel, and Grüne.

One of the truly universal ideas in mathematics these days is that of "exact sequence." Recall that a sequence of mappings

$$A \xrightarrow{f} B \xrightarrow{g} C$$

is *exact* if the image of f equals the kernel of g. From whence does this terminology originate?

It seems that the *concept* of exact sequence originates with Hurewicz (*Bulletin of the American Mathematical Society* 47(1941), 562). But he did not use the term. In fact the terminology made its debut in a paper of John L. Kelley and Everett Pitcher (*Annals of Mathematics* 47(1947), 682—709). Although they may have the first citation in print, they humbly attribute the terminology to Eilenberg and Steenrod, who cooked up the language for use in their classic text *Foundations of Algebraic Topology*. Evidently, in their original draft for that book, they left a blank space for every occurrence of the idea. They were seeking just the right language, and were waiting for inspiration to hit. Eilenberg used the term "exact sequence" in a course he taught at the University of Michigan in 1946. He and Steenrod adopted it for their book, which was published in 1952. In the abstracts of talks that Kelley and Pitcher submitted in 1945 and 1946, they alluded to the idea of exact sequence, but they used the language "natural homomorphism sequence." We can be grateful that, for their published work, they adopted the more elegant language "exact sequence." That is the argot that lives on today.

Orin Frink (1901–1988) was one of the spiritual leaders in the Penn State Math Department in the post-World War II period. He cut a somewhat austere figure in an otherwise rather informal department. One day some of his graduate students knocked on his office door to ask a question about a certain complex series whose convergence they had been assigned to determine. Frink opened the door just about one inch to hear the question, "Sir, we would like to know whether this series converges." His dark eye peering out through the crack, Frink muttered, "Yes," and slammed the door.

One year, when Frink was Chairman of the Department, he took all the graduate student admission letters for the following year, stuck them in his inside jacket pocket, and left the building for the day intending to drop the letters in the mail. But he got distracted and forgot all about the letters. He hung the jacket in his closet at home, and did not take it out to wear again until the following Fall. Too late! No graduate students had shown up to begin the school year.

Per Enflo (1944–) caused quite a sensation in the early 1970s by solving two of the biggest unsolved problems in Banach space theory: he gave a counterexample to the approximation problem, and a counterexample to the invariant subspace problem. He in fact always was a phenomenon. In the same year, as a student, he won both a national piano competition in Sweden and a national mathematics competition. Today he couples his study of mathematics with new studies in anthropology.

Sir Michael Atiyah (1929–) once gave an interview in which he pondered his life, and the course of his career. Of course he is famous for, among other things, contributing to the invention of K-theory and proving the Atiyah-Singer Index Theorem (for which he has received both the Fields Medal and the Abel Prize). He was asked how he selected a problem for study. The quite thoughtful answer was

Michael Atiyah

I think that presupposes an answer. I don't think that's the way to work at all. Some people may sit back and say, 'I want to solve this problem,' and they sit down and say, 'How do I solve this problem?' I don't. I just move around in the mathematical waters, thinking about things, being curious, interested, talking to people, stirring up ideas; things emerge and then I follow them up. Or I see something which connects up with something else I know about, and I try to put them together and things develop. I have practically never started off with any idea of what I'm going to be doing or where it's going to.

...

You can't develop completely new ideas or theories by predicting them in advance. Inherently, they have to emerge by intelligently looking at a collection of problems. But different people must work in different ways. Some people decide that there is a fundamental problem that they want to solve ... I've never done that, partly because that requires a single-minded devotion to one topic which is a tremendous gamble.

...

Now some people are very good at that; I'm not really. My expertise is to skirt the problem, to go round the problem, behind the problem ... and so the problem disappears.

$\mathcal{T}$homas Alva Edison (1847–1936) had in his employ a graduate of Princeton University named Francis Robbins Upton (1852–1921). In fact Upton had studied in Germany under the great scientist Helmholtz (1821–1894). One day Edison wanted to know the volume of the pear-shaped bulb that he had devised for the electric lightbulb. Upton was assigned to the problem, and he labored away with pencil and paper. Long hours passed, and Upton was still at work. After waiting a few days, the great inventor asked the young man for the result. "I've not yet finished. I need more time." Edison became impatient. He blurted, "Let me show you how to do it." He then poured water into the bulb. "Now measure the water and you'll have the answer."

$\mathcal{F}$red Gehring (1925–), a distinguished complex analyst, was Chairman of the Math Department at the University of Michigan for ten years. He ruled with a firm but benevolent hand, and many of his policies live on. One of these is the concept of "Gehring number." If you are offering a graduate course, then your Gehring number is the number of students enrolled plus half the number of auditors. If your Gehring number exceeds five, then you get to teach the course. Otherwise the course is cancelled and you get to teach calculus.

$\mathcal{S}$eymour Cray (1925–1996) was arguably the godfather of supercomputing. For many years, the fastest parallel processing machines came out of the Cray plant outside of Minneapolis. And many of the hottest new high tech companies were spin-offs from Cray. For quite a time, the speed of a Cray was the benchmark for the computing world. As a simple example, when the first Pentium chips came out people crowed that "now we have a Cray I on a chip." That chip ran at 100 megaflops.

Seymour Cray was a special and enigmatic man. When he had a deep and difficult problem to think about, he would go underground. Literally. For many years, Cray worked on digging—with a hand shovel—a $4' \times 8'$ tunnel that was to connect his house with a nearby lake. Cray said he always did his best thinking while he was shoveling away at his tunnel. He died in an automobile accident in Colorado before the tunnel was completed.

Ray Redheffer was for many years one of the grand old men of the UCLA Math Department. He was a body-builder, and would hold his office hours while pumping iron. He did handsprings and back-flips in class (this was all while he was in his sixties!). The students loved him.

Ray's graduate students helped him build his house—they were volunteers, and Ray compensated them handsomely. The house was a real showpiece, obviously a very original piece of work. And, as was the habit of UCLA Professors in the 1960s and 1970s, he had a separate little building in the backyard where he did his mathematics. One day Ray proudly showed me his study: He opened the door—which looked just like any other door to a building—and behind it was another door. He opened that door, revealing a crypt-like space that was about 4 feet wide, 8 feet long, and 7 feet high. There was no furniture (where would one put it?), no light switch, no windows. Ray explained that he would come in here, close both doors, and lie on his back in total darkness and silence. This is where he loved to do his mathematics.

Ray Redheffer is a mathematician of many talents, and diverse interests. In addition to being a vigorous and active scholar, he is a gymnast and weightlifter. He has penned several prominent textbooks. He has written papers on differential equations, inequalities, maximum principles, harmonic analysis, complex analysis, and many other parts of mathematics. One of his more remarkable discoveries is an elementary question about matrices of 0s and 1s that turns out to be equivalent to the Riemann Hypothesis.[*]

---

[*] See R. Redheffer, Eine explizit lösbare Optimierungsaufgabe, Band 3 (Tagung, Math. Forschungsinst., Oberwolfach, 1976), pp. 213–216. *Internat. Ser. Numer. Math.*, v. 36, Birkhäuser, Basel, 1977.

In fact the Redheffer matrix is this: Let $A(n)$ be an $n \times n$ matrix of 0s and 1s defined by

$$A(i, j) = \begin{cases} 1 & \text{if } j = 1 \text{ or } i \text{ divides } j \\ 0 & \text{otherwise} \end{cases}$$

The Riemann hypothesis is true if and only if $\det(A) = \mathcal{O}(n^{1/2 + \varepsilon})$ for each $\varepsilon > 0$.

"I hate fractals. They remind me of something from *Dairy Queen*."
— Howard Nemerov (1920–1991), Poet Laureate of the United States

The Ziwet Lectures at the University of Michigan have had a seminal influence on the development of modern mathematics. In 1941, Saunders Mac Lane delivered this lecture series. During one of these lectures (on group extensions), people were surprised to see Sammy Eilenberg suddenly leave the room in great haste. People of course wondered whether something was wrong; they later learned that Sammy had just then realized the important connection between group extensions and topology. In the following days Eilenberg and MacLane were often found conferring intensely. The ensuing years saw a flood of important collaborative work by these two authors.

Sammy Eilenberg and Saunders Mac Lane

*J*t is always amazing to witness the different ways in which people locate themselves in the world. When I visit my in-laws in South Dakota, I find it charming to hear one farmer explain to another how to find a certain site: "You go to the place where Jeb ran over that dog with his combine last year, turn left, go over the crick, and then drive past Mrs. Swensen's wheat field. When you get to the tree that was struck by lightning in the big storm, go two scootches to the right and you'll be right there. You can't miss it."

By contrast, I recently visited the Math Department at Boston College. At lunch one day, someone asked where a new Thai restaurant was located. The immediate reply was, "It's right by the place where the Red Guard fire-bombed that *New York Times* reporter's car last year."

*J*ulian Lowell Coolidge (1873–1954) was a real aristocrat, and a member of the Harvard math faculty in the 1930s. During class one day, while he twirled his watch on its chain, the chain broke and the fob watch flew out the window. Coolidge exclaimed, "Ah, gentlemen, a perfect parabola." Coolidge was a geometer.

A student of Plato (428 B.C.–348 B.C.) once asked the great master, "What practical use do these theorems serve? What is to be gained from them?" Plato's answer was immediate and peremptory. He turned to one of his slaves and said, "Give this young man an obol [a small Greek coin] that he may feel that he has gained something from my teachings. Then expel him."

*N*orbert Wiener was teaching a graduate class at MIT. One day he walked into the classroom and announced: "Today I am going to prove the Riemann Hypothesis." He began lecturing, providing all the minute details of his proof, and filling up the blackboard with complex and recondite mathematics. This procedure went on for the whole hour and at the end he was not finished. In the next class period he continued where he left off. Finally, on the third day, he was still going strong until he finally wrote down a statement that was clearly incorrect—it contradicted an earlier statement he had made. Wiener stood back and looked at it for a while and then said: "I guess I have proved too far."

*I*t is difficult for the non-expert to understand or appreciate the over-riding joy, and all-consuming passion, with which a mathematician regards his subject. Mathematics is the be-all and end-all of his life.

When André Weil was a child, he had a painful fall. His sister brought him an algebra book to comfort him. It worked. Later in life, Weil said, "Every mathematician worthy of the name has experienced, if only rarely, the state of lucid exaltation in which one thought succeeds another as if miraculously ... Unlike sexual pleasure, this feeling may last for hours at a time, even for days."

When, at the age of eighty-nine, John E. Littlewood had to spend time in a nursing home to recover from a bad fall, a friend (Béla Bollobás, 1943–) tried to cheer him up with a math problem:

> In my desperation I suggested the problem of determining the best constant in Burkholder's weak L_1 inequality. To my immense relief (and amazement), Littlewood became interested in the problem. It seemed that mathematics did help to revive his spirits and he could leave the nursing home a few weeks later. From then on, Littlewood kept up his interest in the weak inequality and worked hard to find suitable constructions to complement an improved upper bound. Unfortunately, we did not have much success so eventually I published the improvement only after Littlewood's death.

*B*ertrand Russell expresses the rapture of pure mathematics as follows:

> The true spirit of delight, the exaltation, the sense of being more than man, which is the touchstone of the highest excellence, is to be found in mathematics as surely as in poetry. What is best in mathematics deserves not merely to be learned as a task, but to be assimilated as a part of daily thought, and brought again and again before the mind with ever-renewed encouragement. Real life is, to most men, a long second-best, a perpetual compromise between the real and the possible; but the world of pure reason knows no compromise, no practical limitations, no barrier to the creative activity embodying in splendid edifices the

Bertrand Russell

passionate aspiration after the perfect from which all great work springs. Remote from human passions, remote even from the pitiful facts of nature, the generations have gradually created an ordered cosmos, where pure thought can dwell as in its natural home, and where one, at least, of our nobler impulses can escape from the dreary exile of the natural world.

$\mathcal{G}$. H. Hardy offers this rejoinder to Russell's remarks: "It is a pity that it should be necessary to make one very serious reservation—he must not be too old. Mathematics is not a contemplative but a creative subject; no one can draw much consolation from it when he has lost the power or desire to create; and that is apt to happen to mathematicians rather soon."

$\mathcal{M}$*athematical Reviews*, now most commonly encountered in the OnLine version `MathSciNet`, had its first issue appear in January, 1940. Fittingly, it celebrated its sixtieth birthday in the year 2000. At that time some interesting data were uncovered:
- Paul Erdős had an article reviewed both in the year 1940 and the year 2000.
- Both H. S. M. Coxeter and Dirk Struik wrote reviews in 1940 and 2000.

$\mathcal{O}$ne of the most remarkable mathematical tracts ever written is G. H. Hardy's *A Mathematician's Apology*. Written after one of Hardy's suicide attempts, the book is both a genuine apology and a note of despair, but also a celebration of the mathematical life. In one passage Hardy says

It is a melancholy experience for a professional mathematician to find himself writing about mathematics. The function of a mathematician is to do something, to prove new theorems, to add to mathematics, and not to talk about what he or other mathematicians have done.

...

If then I find myself writing not mathematics but 'about' mathematics, it is a confession of weakness, for which I may rightly be scorned or pitied by younger and more vigorous mathematicians. I write about mathematics because, like any other mathematician who has passed sixty, I have no longer the freshness of mind, the energy, or the patience to carry on effectively with my proper job.

*T*here has long been some evidence of a correlation between mathematics and left-leaning politics. As an example, student radical leader Mark Rudd (1947–), who during the 1960s was an organizer of SDS (Students for a Democratic Society) and who led the student foment at Columbia University, today teaches junior college mathematics in Albuquerque, New Mexico. Rudd was one of the more aggressive national student leaders in the protests against the Vietnam War. His group occupied the Columbia University President's office and actually held the Dean hostage. Rudd later helped to found the considerably more radical (and much more violent) spinoff group, the Weathermen. He is the man who coined the phrase, "Up against the wall, motherfu–." Today, Rudd wonders what got into him thirty-five years ago.

*S*wedish mathematician Bjorn Dahlberg (1949–1998) (pronounced *Dalberry*) was on the Math Faculty at Washington University when he got involved in a project with the Volvo Corporation. Describing Volvo's "mathematical problem" requires a bit of background.

Those of us who saw filmstrips about it in grade school may recall how automobile bodies are (traditionally) designed:

- The artists dream up a new body shape, and render it in clay.
- The artists show their product to the engineers, who critique it from the point of view of engine design, passenger space, aerodynamics, and so forth.
- The artists return to the drawing board and create a new design. They render it in clay.
- The engineers have another go at criticizing the new clay model.

and so forth. The process just described would often take 1000 man-hours or more in the path from initial inspiration to a working design.

The challenge that Dahlberg faced was to find a way to carry out this surface-design process on a computer. The "math team" at Volvo had looked at the problem and declared it to be intractable. Such nay-saying had never discouraged Bjorn Dahlberg. Volvo really wanted this problem solved, and they were willing to throw lots of money at it. That didn't hurt Dahlberg's feelings a bit.

Dahlberg led a team of scientists from Gothenberg University and from Volvo to tackle the problem. They realized early on that an automobile body is the disjoint union of convex surfaces. Their concept is that the designer can specify a convex surface by choosing points in space through which it will pass (or nearly pass) and also specifying attributes for those points: light refraction, aerodynamic factors, and so forth. The computer solves an optimization problem—in fact using a variant of Karmarkar's ideas on linear programming—and produces a surface.

The software product that they produced, called *SLIP*, has revolutionized the design of automobile bodies. It also has applications in many different areas of endeavor that involve surface design. At the time of Dahlberg's untimely death a few years ago, he was engaged in discussions with the America's Cup racers.

$\mathcal{A}$lfred Tarski was one of the leading logicians of the twentieth century. A professor at U. C. Berkeley for many years, he exerted a powerful influence over many students and colleagues. Tarski preferred to work at night, and he slept late into the morning to compensate. His evening regimen went something like this: First there was dinner, to which his assistant was invited. This repast would include several glasses of buffalo-grass flavored vodka (the Russian style) followed by a heavy meal with lots of wine. After dinner, Tarski and his collaborator would repair to the study on the first floor of the house in Berkeley. Apart from the general lethargy induced by the liquor and heavy food, there was a great distraction with the panoramic view of the large garden and of San Francisco Bay. Nevertheless, Tarski was determined to work. He would light up a cigarillo, and they would begin.

Sometimes, around midnight, Tarski would require some stimulant. Often it was *Kola Astier*, an exotic potion (not available in this country) that contained caffiene and perhaps some cocaine. By midnight, Tarski was flying and his assistant could usually hardly keep his eyes open. By 2:00AM or 3:00AM in the morning, Tarski would be mumbling under his breath about the lack of stamina in the younger generation. By 4:00AM they would usually call it quits.

*A*lfred Tarski's first academic passion was for biology. He entered the university with the intention of studying that subject. But he got derailed by an untimely success in mathematics. To wit, he was taking a course from Lesniewski in set theory, and the professor mentioned a particular open problem to the class. Tarski was able to solve it, and ended up publishing a paper. Tarski later said that he thought the paper not particularly interesting; but at the time the young savant was intoxicated by this success. Egged on by his professor, Tarski changed disciplines and became a mathematician. Later in life, he often expressed regret at having abandoned his first love. He said that perhaps some day he would be reincarnated as a biologist.

*S*ol Fefferman is enthusiastic in describing Alfred Tarski's "superhuman" mathematical attributes, but observes that there was a rather simple humanity about the man. He frequently lectured in U. C. Berkeley's Dwinelle Hall, many of whose rooms had rather small podiums. Tarski always seemed to have an uncertain relationship with material objects and he was constantly seen backing into or rebounding off of some podium. The audience was always afraid of what would happen. Although he frequently teetered, he never fell. And often, because of the forcefulness with which he wrote on the blackboard, the chalk would explode in his hand. And then there were the cigarettes. Tarski was an inveterate chain smoker. He smoked constantly, particularly while he was lecturing. There was this constant interplay of the cigarette and the chalk. It often looked as though he would smoke the chalk and write with the cigarette. But in fact he never did; somehow nature kept the two cylinders focused on their different tasks.

*L*eonard Eugene Dickson (1874–1954), a native Texan, was one of the great mathematical talents of his day. He agreed to a three-year appointment as Associate Professor at the University of Texas beginning in the summer of 1899. But then he got an offer from the University of Chicago, so he decided to abandon Texas and go to Chicago in April of 1900. This action was taken by some regents and faculty members in Texas as at least a gross insult if not morally and legally wrong. Regent Cowart (1845–1924), from his Washington, DC office, wrote to the chairman of the board, T. S. Henderson (1859–1937), that he was "disgusted" with Dickson's resignation. "It seems we are fated to develop men of extraordinary mathematical genius, and then other institutions of learning appropriate them as soon as they show any promise." Cowart also wrote to Proctor Clark (1831–1908): "I am simply disgusted with the infernal way that some of our professors have of coming before us constantly and clamoring for an increase in salary, and when any prize glitters before their eyes in a northern University they quit us without any excuse. I think Dickson's conduct is simply infamous, and I intend, if I can, to have a set of resolutions adopted characterizing it as it should be."

The regents had their revenge on Dickson by not paying him for the last two months he taught—the remainder of the spring term. Dickson fought back by hiring a lawyer from his hometown of Cleburne, Texas, but the records do not show whether he ever prevailed.... T. M. Putnam (1875–1942), who came to Texas as an instructor from California, apparently with Dickson, left with him for Chicago and was one of his first doctoral students there. During Dickson's last brief tenure in Texas, he was R. L. Moore's calculus instructor.

Dickson is remembered today as the namesake of the endowed Dickson Instructorships at the University of Chicago.

$\mathcal{L}$. E. Dickson was a top scholar of the University of Chicago Math Department in the 1920s and 1930s. He was highly regarded and remembered fondly by colleagues and students alike. W. L. Duren, Jr. (1905–) recalls him in this way:

> In the conventional sense Dickson was not much of a teacher. I think his students learned from him by emulating him as a research mathematician more than being taught by him. Moreover, he took them to the frontier of research, for the subject matter of his courses was usually new mathematics in the making. As Antoinette Huston said, "He made you want to be with him intellectually. When you are young, reaching for the stars, that is what it is all about." He was good to his students, kept his promises to them and backed them up. Yet he could be a terror. He would sometimes fly into a rage at the department bridge games, which he appeared to take seriously. And he was relentless when he smelled blood in the oral examination of some hapless, cringing victim. He was an indefatigable worker and in public a great showman, with the flair of a rough and ready Texan. An enduring bit in the legend is his blurt: "Thank God that number theory is unsullied by any application." He liked to repeat it himself as

well as his account of his and his wife's honeymoon, which he said was a success, except that he got only two papers written.

*I*t is not generally well known that Alan Turing (1912–1954), of Turing machine fame, actually designed a machine to calculate the zeros of the Riemann zeta function. According to Andrew Hodges (1949–), the Turing biographer,

> Apparently [Turing] had decided that the Riemann Hypothesis was probably false, if only because such great efforts have failed to prove it. Its falsity would mean that the zeta function did take the value zero at some point which was off the special line, in which case this point could be located by brute force, just by calculating enough values of the zeta function.

Turing did his own engineering work, hence he got involved in all the fine details of constructing this machine. He planned on eighty meshing gearwheels with weights attached at specific distances from their centers. The different moments of inertia would contribute different factors to the calculation, and the result would be the location of and an enumeration of the zeros of ζ.

Visitors to Turing's apartment would be greeted by heaps of gear wheels and axles and other junk strewn about the place. Although Turing got a good start cutting the gears and getting ready to assemble the machine, more pressing events (such as World War II) interrupted his efforts. His untimely death prevented the completion of the project.

*P*eter Hilton (1923–) tells of working on British cryptography efforts during World War II. In particular, he worked with Alan Turing on deciphering the German Fish Machine. As he tells it,

> In general, we had to rely on small statistical biases in the German language to eliminate most of the myriad possibilities; and then we used "hand methods" to make the final determinations. It was Alan Turing who first appreciated the essential role which could be played in the elimination phase of the process by high-speed electronic machines, and who was, in fact,—and quite consciously and deliberately—inventing the computer as he designed the first "Bombe" and then the "Colossus" for our cryptanalytical purposes.

Peter Hilton describes Turing as a superb athlete. For example, Turing was a marathon runner of considerable distinction. Hilton tells of Turing's particular prowess at playing doubles tennis. Turing was extremely quick at intercepting attempted passing shots; but he became quite frustrated at the frequency with which he put the ball into the net. Turing applied deep thought to the solution of this problem, and he decided that it was necessary to increase the time interval between the

Alan Turing

arrival of the ball at his racquet and the ball's subsequent return. The remedy that he then devised was to loosen the racquet strings. He performed this job himself, and he loosened the strings considerably.

The next time that Turing arrived on the tennis court, his racquet bore a distinct resemblance to a fishing net or a butterfly net. The racquet had virtually no good use in the game of tennis, and it can arguably be said that Turing's performance was considerably worse than before (but for different reasons!). He did finally admit that the alterations to his racquet were not in the spirit of the game.

Jt is said that Wiener was the ultimate intuitive mathematician. His flights of fancy were both inspiring and bewildering.

But von Neumann did everything analytically and logically. He loved fussy detail, needless repetition, and explicit notation. In one paper he developed an argot that is reminiscent of the computer language *LISP*. Namely, he defined a generalization of the usual "function" notation that he denoted $f((x))$. Later in the paper he had to refine the idea, so he introduced $f(((x)))$. Still later, another refinement was introduced. So now we have $f((((x))))$. Equations such as

$$(\psi((((a)))))^2 = \phi((((a))))$$

have to be peeled before they can be digested (Paul Halmos's words). Some have referred to this paper as "von Neumann's onion."

$\mathcal{E}$ven great mathematicians can have a rocky start on the road to mathematical eminence. Stories about Einstein's (1879–1955) struggles with mathematics abound. Less well known is John E. Littlewood's travails:

> ... I had a happy childhood among mountains, the ocean, and a beautiful climate. Education was inevitably meager—adequate staff in a Colony was then very hard to get. I remember failing in arithmetic (my only failure in examinations), and that algebra as taught was quite unintelligible. I had a period at the University of Cape Town, but my father soon realized that I was doing no good there, and I left the University at age 14 to go to St. Paul's School in England.

$\mathcal{J}$ean Dieudonné (1906–1992) waxes euphoric in describing how Bourbaki cuts to the quick of mathematics:

> What remains: the archiclassic structures (I don't speak of sets, of course), linear and multilinear algebra, a little general topology (the least possible), a little topological vector spaces (as little as possible), homological algebra, commutative algebra, non-commutative algebra, Lie groups, integration, differentiable manifolds, Riemannian geometry, differential topology, harmonic analysis and its prolongations, ordinary and partial differential equations, group representations in general, and in its widest sense, analytic geometry. (Here of course I mean in the sense of Serre, the only tolerable sense. It is absolutely intolerable to use *analytic geometry* for linear algebra with coordinates, still called analytical geometry in the elementary books. Analytical geometry in this sense has never existed. There are only people who do linear algebra badly, by taking coordinates and this they call analytical geometry. Out with them! Everyone knows that analytical geometry is the theory of analytical spaces, one of the deepest and most difficult theories of all mathematics.) Algebraic geometry, its twin sister, is also included, and finally the theory of algebraic numbers.

$\mathcal{T}$he Tripos (the advancement exam at Cambridge—see *Mathematical Apocrypha*) became such a dominant institution at Cambridge that everyone focused their undergraduate years on studying for the exam. There grew a cadre of professional coaches for the Tripos. These men were quite celebrated, and the ones who had had many successful students were held in high regard and paid handsomely. The training under the coaches for the Tripos lasted ten terms, and the student typically took the exam early in January of the fourth year of residence. Such distinguished scholars as E. J. Routh (1831–1907, who also wrote some important textbooks), R. R. Webb (1850–1936), and W. H. Young (1863–1942) served as Tripos coaches.

Students in training for the Tripos were taught to be able to state and prove any theorem on the syllabus, and also to solve any problem, at lightning speed. Thus it was not enough just to be a good mathematician. One also had to be very quick. Arthur Cayley (1821–1895) himself took the Tripos in 1842. At that time the exam—which extended over several days— had in it a two-hour paper consisting only of problems. Candidates were not expected to do all the problems; they had a choice. On the evening after this problem paper, a friend called on Cayley to inquire how things went. He said to Cayley, "I've just seen Smith (Smith was one of Cayley's rivals). You know, he did all the questions within two hours." Cayley replied, "Well, I cleaned up that paper in forty-five minutes."

When Lord Kelvin (1824–1907) was a student he still went by his given name of William Thomson of Peterhouse. He was widely regarded to be the most capable mathematics student of his year. Everyone expected him to be a Senior Wrangler (i.e., the top scorer on the Tripos). On the day that the results of the Tripos were published, Kelvin told his college servant, "Go down to the Senate House and see who is Second Wrangler." A short while later the servant returned and said, "*You*, sir, are the Second Wrangler." Evidently one of the other students was quicker than Kelvin.

*O*f course Arthur Cayley is remembered as one of the great heroes of late nineteenth century British mathematics. In those days, as a reaction to the excesses of the Newtonians in their struggle with the Europeans over the primacy to calculus, applied mathematics ruled the roost in Britain and pure mathematics was in remission. Cayley was one of the standout pure mathematicians. James Clerk Maxwell (1831–1879) described Cayley in these words:

> Whose soul, too large for ordinary space,
> In n dimensions flourished unrestricted.

Even the college manciple at Cambridge, a true old-world type among college servants, paid due homage to Cayley. When expounding the glories of Trinity College to visitors, he was known to say, "he's that exact, he could take the earth in his hand and tell you its weight to a pound."

$\mathcal{N}$elson Dunford (1906–1986) was one of the leading lights in functional analysis at Yale during the post-War period. He and Jacob Schwartz spent twenty years writing the epic three-volume book *Linear Operators*. Dunford had quite an unusual youth. He grew up in the depression. He was passed over for a graduate fellowship in the 1930s, and spent a couple of years surviving on $10 a month while studying in the St. Louis public library. Amazingly, the library subscribed to the few mathematics research journals that existed in those days. And Dunford read them. Dunford managed to write his first research paper, dealing with integration of functions taking values in a Banach space, while studying on his own. After the paper was accepted in the *Transactions of the AMS*, Dunford was offered an Assistantship at Brown University. There he earned his PhD under the direction of J. D. Tamarkin (1888–1945). Dunford was hired at Yale right after receiving his degree, and spent his entire career there.

$\mathcal{F}$or most of the twentieth century, mathematicians and physicists asked themselves whether God plays dice with the universe. The question has yet to be answered. Albert Einstein and Niels Bohr were among the first to contribute to the discussion—in the context of considering the validity of quantum mechanics. More recently Stephen Hawking (1942–) has added his own wisdom. The accrued sophistry is as follows:

> "God does not play dice with the universe."   — Albert Einstein

> "Who are you to tell God what to do?"   — Niels Bohr

> "God not only plays dice, but sometimes throws them where they cannot be seen."   — Stephen Hawking

$\mathcal{P}$eter Hilton tells of the world's first programmer: "… I would be totally remiss were I not to mention the world's first programmer, Augusta Ada Byron (1815–1852)." She was Lord Byron's (1788–1824) only child and was separated from him when she was but a month old, at which time he left England forever. Each of them died at age thirty-six, and they lie together in the Byron vault in Nottinghamshire. Byron had great affection for her and wrote of her in Canto III of his *Childe Harold's Pilgrimage:*

> My daughter! with thy name this song begun—
> My daughter! with thy name thus much shall end—
> I see thee not,—I hear thee not,—but none
> Can be so wrapt in thee; thou art the friend
> To whom the shadows of far years extend:
> Albeit my brow thus never shouldst behold,
> My voice shall with thy future visions blend,
> And reach into thy heart,—when mine is cold,—
> A token and a tone even from thy father's mould.

In any case, as the Countess of Lovelace, she did very much for Charles Babbage (1791–1871) and among other things programmed the calculation of the so-called Bernoulli numbers for his never-to-be-finished computer. Her description of Babbage's machine is perhaps worth repeating:

> We may say most aptly that the Analytical Engine weaves *algebraical patterns* just as the Jacquard-loom weaves flowers and leaves. Here, it seems to us, resides much more of originality than the Difference Engine be fairly entitled to claim.

*S*tarting in winter of 1894, several women came to Göttingen to study with Felix Klein (1849–1925). Grace Chisholm (1868–1944), who later married the English analyst W. H. Young, was said to have been Klein's favorite student. She wrote home with an account of the first day of classes that year:

Grace Chisholm Young

> Klein had his first lecture on the hypergeometric functions .... Miss Winston and I made for the Sanctum and found Klein there working till lecture time. Klein, instead of begin-

ning with his usual "Gentlemen!" began "Listeners!" ["*Meine Zuhörer*"] with a quaint smile; he forgot once or twice and dropped into "Gentlemen!" again, but afterwards he corrected himself with another smile. He has the frankest, pleasantest smile and his whole face lights up with it. He spoke very slowly and distinctly and used the blackboard very judiciously. Mr. Woods said he never heard anyone lecture so well and neither have I. I found my notes afterwards perfectly clear though queerly spelt; but I understood as well as at an English lecture.

One might wonder how the important journal *Mathematical Reviews* came about. After all, the German *Zentralblatt für Mathematik* was already well-established at the time (1940), and creating an archival review journal is an expensive and laborious process. Ivan Niven tells the story as follows:

> … the German reviewing journal, *Zentralblatt für Mathematik*, began to deteriorate in the midthirties because Nazi racist policies were being forced on the journal, so that, for instance, the reviewing staff should be "racially pure." This trend led to the decision here to establish a reviewing journal in this country, although there was concern whether American mathematicians had adequate financial support and general mathematical strength to launch such a journal properly. Enter Otto Neugebauer (1899–1990), who came to Brown University in 1939, having played a key role in the founding of *Zentralblatt*. His knowledge and experience were of crucial importance in getting *Mathematical Reviews* started off on the right foot.

James Gleick (1954–) was an English major at Harvard. After graduation he and some associates endeavored to start up a newspaper in Wisconsin. It failed. He ended up as a reporter for the New York Times. With considerable insight and hard work, Gleick agglomerated what he learned from a great many interviews and wrote the book *Chaos*. This was a remarkable effort, and it garnered Gleick considerable fame and fortune.

His next book *Genius*, a biography of Richard Feynman, was also noteworthy. The first account of the life of that great physicist, it is certainly a worthwhile read.

Gleick's newest book is a biography of Isaac Newton. He was recently interviewed on National Public Radio by Diane Rehm (1936–) about this effort. The conversation went swimmingly—up to a point. Gleick was discussing the great controversy between Newton and Leibniz about the provenance (and proper accreditation) of the discovery of the calculus.

Rehm interrupted Gleick with, "But what about trigonometry?" Gleick, not a mathematician by training, cleared his throat and said, "That really doesn't have anything to do with calculus." And he resumed his story.

But Rehm would have none of it. She interrupted again and said, "Yes, but *what about trigonometry?*" One could almost hear Gleick's teeth grinding over the radio waves.

*T*here is considerable concern these days for engaging students, especially freshmen and sophomores, in the learning process of mathematics. The reform movement has declared that "The lecture is dead," and for more than fifteen years we have been wrestling with that declaration. Little do most of us know that our vituperations were anticipated by Solomon Lefschetz many years ago:

> I have found that in freshman courses in mathematics, and less so in the next year, hardly one third of the students care for and are not totally bored by mathematics. Hence at that early level a teacher must be exceptionally lively and have a sympathetic understanding of the students. Needless to say this must be coupled with a complete grasp of the topic taught.
>
> Here are a few very radical suggestions for later years. From the junior year on through graduate work they should be merged into a professional school, with teaching, at least in mathematics, of seminar type plus abundant but easy contact with faculty on an individual basis. In other words, "baby talk" should end with sophomore years.

*I*n *Mathematical Apocrypha*, we noted several celebrities with a background in mathematics. A new addition to this list is comic Steve Martin, who studied math in school. He was recently featured in a public forum with Robert Osserman, Associate Director of the Mathematical Sciences Research Institute in Berkeley, California. Osserman is a very distinguished mathematician, formerly of Stanford University. Seems that Osserman played straight man to Martin's quips. Osserman has become quite the celebrity at MSRI. On another occasion he did a two-man show with playwright Tom Stoppard. His most recent encounter has been with writer Rebecca Goldstein.

There was concern in the 1930s that the influx of distinguished scholars from Europe would use up the few available jobs, thus in effect punishing those young Americans who were trying to get into the academic profession. Ivan Niven tells it this way: "Although the emigrés were generally received with friendship and cordiality, there was one unhappy incident. In the late thirties, Richard Courant (1888–1972) of New York University was invited to Yale University to speak. It fell to Einar Hille (1894–1980) 'painfully to disinvite him,' not because of antisemitism, but rather xenophobia among the graduate students who felt keenly the 'unfair,' as they saw it, competition of outstanding foreign mathematicians for the few available jobs." This quotation is taken from a letter dated September 23, 1987, to Peter Lax (1926–) of NYU from Asger Aaboe (1922–), Professor of the History of Science at Yale.

When Harvard mathematician Benjamin Peirce (1809–1880) heard that the Johns Hopkins University was to be founded in Baltimore, he wrote to the new president Daniel Coit Gilman (1831–1908) in 1875 as follows:

> Hearing that you are in England, I take the liberty to write you concerning an appointment in your new university, which I think would be greatly for the benefit of our country and of American science if you could make it. It is that of one of the two greatest geometers of England, J. J. Sylvester. If you enquire about him, you will hear his genius universally recognized but his power of teaching will probably be said to be quite deficient. Now there is no man living who is more luminary in his language, to those who have the capacity to comprehend him than Sylvester, provided the hearer is in a lucid interval. But as the barn yard fowl cannot understand the flight of the eagle, so it is the eaglet only who will be nourished by his instruction …. Among your pupils, sooner or later, there must be one, who has a genius for geometry. He will be Sylvester's special pupil—the one pupil who will derive from his master, knowledge and enthusiasm—and that one pupil will give more reputation to your institution than the ten thousand, who will complain of the obscurity of Sylvester, and for whom you will provide another class of teachers …. I hope that you will find it in your heart to do for Sylvester—what his own country has failed to do—place him where he belongs—and the time will come, when all the world will applaud the wisdom of your selection.

$\mathcal{H}$. F. Baker (1866–1956) gives the following history of J. J. Sylvester and the Johns Hopkins University:

> In 1875 the Johns Hopkins University was founded at Baltimore. A letter to Sylvester from the celebrated Joseph Henry (1797–1878), of date 25 August 1875, seems to indicate that Sylvester had expressed at least a willingness to share in forming the tone of the young university; the authorities seem to have felt that a Professor of Mathematics and a Professor of Classics could inaugurate the work of a University without expensive buildings or elaborate apparatus. It was finally agreed that Sylvester should go, securing besides his travelling expenses, an annual stipend of 5000 dollars "paid in gold." And so, at the age of sixty-one, still full of fire and enthusiasm, ... he again crossed the Atlantic, and did not relinquish the post for eight years, until 1883. It was an experiment in educational method; Sylvester was free to teach whatever he wished in the way he thought best; so far as one can judge from the records, if the object of an University be to light a fire of intellectual interests, it was a triumphant success. His foibles no doubt caused amusement, his faults as a systematic lecturer must have been a sore grief to the students who hoped to carry away note-books of balanced records for future use; but the moral effect of such earnestness ... must have been enormous.

Utter Sanguinity

$\mathcal{M}$arilyn vos Savant (1946–) writes a popular, syndicated newspaper column called "Ask Marilyn." ["Marilyn vos Savant" is not her real name. In fact "vos Savant" is her mother's maiden name.] Ms. vos Savant is reputed to have the highest IQ in history, and that is the basis for her celebrity and her column. She is very clever, and has a knack for answering difficult questions (she has the good sense to ask experts when confronted with a problem on which she has no expertise). She rarely makes a mistake, although there is a Web site called *Marilyn is Wrong!* [http://www.wiskit.com/marilyn.html] which claims to point out a number of her *faux pas*.

Actually Marilyn vos Savant was educated in St. Louis (my stamping ground). She attended Meramec Junior College, but did not graduate. She also attended Washington University (my institution), but did not graduate. She is married to Robert K. Jarvik (who invented the artificial heart), and makes her home in New York City.

Marilyn vos Savant gained particular celebrity when one of her readers wrote in to ask about what has become known as the "Monte Hall Problem." The question is modeled after the activities that host Monte Hall conducts on his television game show "Let's Make a Deal." Here is the question:

> You are a contestant on the show. You face the stage, on which there are three doors. Behind one door is a Cadillac, and behind the other two doors are goats. You are to choose a door, and you get as your prize whatever is behind that door. You choose door #2, and wait to see what is behind it.
>
> While you are waiting, Host Monte Hall teases you by commanding that his assistant open door #1, revealing a bleating goat standing there. Then he asks you, "OK, one of the remaining doors—#2 or

#3—has a goat behind it and one has the more desirable Cadillac. You have chosen door #2. Based on what I have just shown you, would you like to change your pick to door #3?"

Now a naive analysis of this situation might suggest that the odds that the car is behind door #2 are equal to the odds that the car is behind door #3. So it really does not matter whether you switch or not. But a more careful consideration (especially considering that the sample space is *not* single doors but something more subtle) reveals that if you switch then the odds are two-to-one in your favor to get the car. So you should switch.

Ms. vos Savant published this solution—the correct solution—in her column. Woe is us!! Over two thousand academic mathematicians wrote to Marilyn vos Savant and told her that she was in error. And some of them were not too polite about it. This was quite a debacle for American mathematics.

I have to say that this whole set of circumstances went to Marilyn vos Savant's head. When Andrew Wiles published his solution of Fermat's last problem, Ms. Savant published a little book [SAV] claiming that his solution is incorrect. The basis for her daring allegation is twofold: **(i)** she offers a proof that the complex numbers do not exist (therefore nullifying Wiles's use of said number system) and **(ii)** she observes that Wiles uses hyperbolic geometry, in which the circle can be squared—but everyone knows that the circle cannot be squared—and this is a contradiction.

I wrote to Ms. vos Savant, and to her publisher, pointing out the error of her ways. And she answered! She said, "My mathematician friends and I had a good laugh over your letter. Keep those cards and letters coming." So much for scholarly discourse.

Martin Gardner is particularly chagrined for having been thanked for checking the math in vos Savant's book. He asserts vehemently that he did not. Barry Mazur was also asked to check the math, but he managed to dodge the bullet.

*P*aul Erdős was fanatically attached to his mother. He traveled with her whenever he could, especially when she was older. She tried hard to learn English, and they could often be seen conversing in this new tongue. She frequently turned to Paul with queries about the English language. As an instance, "Palkó, how do you call the fruit 'szilva' in English?" The immediate reply was, "Plimm, mother, plimm!"

Paul Erdős and Paul Turan (1910–1976) had a fondness for poetry, and also for light verse. They composed some of their own, especially concerning the vicissitudes of old age. A favorite was

> One thought disturbs me, that I may decease
> In slowly progressing Alzheimer's disease.

$\mathcal{C}$harlie Fefferman was a plenary speaker at the International Congress of Mathematicians in Vancouver in 1974. He was of course asked to write a paper for the *Proceedings*. The editors rejected his first effort, offering a number of trite and officious criticisms. [I should note that this paper is a masterpiece—one of the most important writings in modern harmonic analysis.] Charlie carefully followed the criticisms and made revisions. They rejected it again. So he lost his cool and wrote them a letter saying that they could publish a blank page with his name on it, saying that Professor Fefferman's remarks will appear elsewhere. Then they relented. Charlie's remarks now appear in the official *Proceedings*.

$\mathcal{E}$mil Artin (1898–1962) was not a great fan of statistics, as the following quotation (1947) illustrates:

> Everybody knows that probability and statistics are the same thing, and statistics is nothing but correlation. Now the correlation is just the cosine of an angle, thus all is trivial.

$\mathcal{T}$hatcher, Arizona is evidently a rather conservative community. A headmistress at one of the local schools decided to remove all the controversial books from the library. When she was finished with her local version of the inquisition, nothing remained on the shelves but mathematics textbooks.

$\mathcal{A}$rmand Borel (1923–2003) was born in La Chaux-de-Fonds, Switzerland. He made an immediate splash on the mathematical scene when he published his thesis on the topology of Lie groups and their classifying spaces in the venerable *Annals of Mathematics*. He became a full Professor at the ETH in Zürich, his *alma mater*, in 1955. He moved on to a permanent membership at the Institute for Advanced Study in Princeton in 1957. He spent the remainder of his life in Princeton.

On the initiative of Georges de Rham, the Swiss federal authorities tried for years to get Borel to return to his native land. Borel's response was, "If Barry Goldwater becomes president, then maybe."

$\mathcal{B}$orel had a favorite story about the concept of "hidden authorship." It concerns the result known as "Chevalley's theorem." Borel heard Claude Chevalley (1909–1984) lec-ture on the result, expressed great

Armand Borel

interest, and asked whether there was a version in characteristic p. Chevalley offered no answer, but Borel kept pestering because he needed the modified result. Chevalley finally gave Borel his notes, indicated that he thought the result too trivial for publication, and declaimed, "Do whatever you want with these!"

Well, Armand Borel was no slouch. He worked up the notes into a full-blown paper, including the characteristic p case. He listed Chevalley as the author, added a footnote that Armand Borel would be using the characteris-tic p result in a forthcoming paper, and submitted the paper to André Weil for publication in the *American Journal of Mathematics*. Since Chevalley was an associate editor for the journal, there was some danger that he would cause trouble. But Chevalley rarely read his mail. So the paper floated through the editorial pipeline and was published.

When the bound issue of the *American Journal* landed on Chevalley's desk, he immediately noticed the paper. In an outrage, he stormed into Borel's office and demanded an explanation. "You told me to do whatever I wanted with your notes," murmured Borel, "and that is what I did with them." Chevalley realized that he had lost the argument and he stormed out, slamming the door behind him.

$\mathcal{T}$he television show *The Simpsons*, in the "Treehouse of Horror" episode from the sixth season, revealed the following counterexample to Fermat's Last Theorem:

Take your TI-83 calculator and compute
$$[1782^{12} + 1841^{12}]^{1/12}.$$
You will find the answer to be 1922. Thus
$$1782^{12} + 1841^{12} = 1922^{12}.$$

We leave it as an exercise for you to make peace between this example and Andrew Wiles's celebrated proof (*Hint:* Think about roundoff error).

$\mathcal{A}$ favorite pastime at the Institute for Advanced Study in Princeton is walking on the grounds. They are, as well as a mathematical preserve, also a game preserve; the flora are abundant and lovely as well. Lie grouper Harish-Chandra and physicist Freeman Dyson were walking and talking one day when Harish-Chandra announced, "I am leaving physics for mathematics. I find physics messy, unrigorous, elusive." Dyson replied, "I am leaving mathematics for the exactly the same reasons." And that is what they both did.

$\mathcal{A}$lfred Tarski liked to socialize, and he was a great host. He was invited to many parties as well. When he was on his own in Berkeley in 1943 (his family had been stranded by the war in Europe), he was invited to a departmental picnic. Everyone was to bring something, and to make things easy for the struggling Polish bachelor, they told Tarski to bring napkins. Naturally Tarski forgot all about it, but in the car on the way to the picnic he told the driver to stop at a drugstore so he could carry out his obligation. He went inside and asked the druggist for napkins. "You mean sanitary napkins," said the druggist. "Of course, sanitary," said Tarski. "A big box?" "Certainly, the biggest." So Alfred Tarski showed up at the picnic with a giant-sized box of Kotex.

$\mathcal{V}$ictor Miller, an old habitué of Harvard, recalls this tale of the grand old life in Cambridge:

> In the summer of 1972 John Tate gave a daily seminar about Elliptic Curves. This is where I really learned about the subject and felt that my notes were irreplaceable until Silverman's book on Elliptic Curves came out. The seminar was held in the one seminar room in 2 Divinity every day starting at 11am.

One day when I arrived, I saw a crowd around the window over-looking the parking lot. Naturally I asked what was going on. They said that a car was being towed away (the parking lot had room for only 5 cars!). It was a black Triumph convertible that I recognized as looking just like one that Raoul Bott had bought the year before (Bott had not been shy about telling anyone who would listen about his "marvelous new car"). I asked if Bott—who had been on leave and was going to be taking over as chairman from Tate—was around. I was told that he was in Tate's office.

I went to the office and knocked on the door—receiving a rather gruff "what is it" from Tate. I opened the door and asked Bott if his car was parked in back of the building. He said that it was, and I told him that it was being towed away. He started groaning and moaning about it and said "If something's not done, I'm resigning as chairman!."

Mari Wilson, the department secretary then made a few calls, and, much to my amazement, the tow truck arrived about 10 minutes later towing the car and put it back into its old spot! To me that really demonstrated the power of a department chairman at Harvard.

*W*hen I was a visitor at the Université de Paris-Sud in Orsay, France, it was always a treat to go into the departmental office. One generally found the secretaries leaned back in their chairs, chain-smoking Gauloises and talking trash in very sonorous, low-key, gutteral French. But as soon as they spied a faculty member—even a humble one like myself—they switched gears into a high-pitched chatter and scurried about in an effort to be of service. I was only an Assistant Professor at the time, but I was graciously offered the amenities of the department in the key of high C. My host, Professor Varopoulos, was a distinguished Professor—holder of the Salem Prize, speaker at the International Congress, and all that—and his needs were met to the accompaniment of a hemisemidemiquaver.

*O*ne day I was in Varopolous's office endeavoring to talk some mathematics. The phone rang. Varopoulos was an impatient man by nature, and he was not happy at this interruption. He picked up the phone with a scowl, and out came the hemidemisemiquaver of one of the staff. Varopoulos scowled even more intensely. "D'accord," he said. The hemisemidemiquaver continued to spew out of the phone. "Oui, oui," he said. Then he put the phone down on his desk. The high pitched jabber continued to emanate from the apparatus. Varopoulos smiled and said, "Let's get back to our mathematics." Somewhat non-plussed, I endeavored to comply. But the phone continued to jabber. After several minutes, Varopoulos picked up the phone and barked, "C'est ça." Then he put down the phone and continued to talk about the unit ball in $\mathbb{C}^n$. The hemisemidemis still poured out of the phone. After five more minutes he picked up the phone once more and said, "Toujours." Then he replaced it on the desk.

This scenario repeated itself for about twenty minutes. Then Varopoulos cleared his throat, adopted a business-like demeanor, grabbed the telephone receiver, and growled, "Maintenant, écoutez…" And he proceeded to tell the secretary in no uncertain terms what he wanted. Then he hung up the phone, gave me a huge smile, and we resumed our mathematics.

$\mathcal{A}$ pending colloquium guest at Rice University was making arrangements for his visit with John Polking of their Math Department. Polking explained that the university had a special deal with a very fancy hotel. He went into some detail to describe the lavish features of the hotel room—actually it was a *suite*—and all the amenities that were available. The guest was suitably impressed. Polking concluded by saying, "The only complaint that we've ever had is that the second bathroom has no telephone."

$\mathcal{M}$y real analysis teacher—a man who had enormous influence on my life—was a very unusual man. His name was Stanton Philipp (1934–1993). He learned much of his basic mathematics while in jail. This is because he disobeyed orders while in the army and was thrown in the brig. Like André Weil, he found that he had a lot of time on his hands, the food was OK and the duties few, and he devoted the many idle hours to learning mathematics. He was a smart guy, and he learned his subject well. But he never heard anyone pronounce the words, and he got lots of them wrong. In class, even in 1970, he spoke of the der͞ivative. And he said "the integral from b to a" to describe $\int_a^b dx$. He was a marvelous teacher, but his approach to everything was eccentric to say the least. He always told me that he wanted a beanie with a propeller, and I just missed sending him one before he died.

*T*atum O'Neal (1963–) is the daughter of actor Ryan O'Neal, and has herself been a movie star since childhood.

At the age of fourteen she was making the movie *National Velvet* (a remake of the original starring Elizabeth Taylor) and suffered a visitation on the set from a school officer who was concerned that she was falling behind in her school work. He asked whether her poor grades in math were not the cause for some concern. "Oh no," she replied confidently. "I have an accountant."

*A*s the sun eclipses the stars by his brilliance, so the man of knowledge will eclipse the fame of others in assemblies of the people if he proposes algebraic problems, and still more if he solves them.

—Brahmagupta, 625 A.D.

*T*he calculus text of Thomas and Finney has been one of the most successful of modern times. One of the recent editions was produced by the editors in some haste, and many of the people involved in production were inexperienced or lacking in judgment. The end result was less than perfect. The ink bled, the graphics were substandard, and the text was riddled with errors. Everyone knows that there is hardly anything more aggravating than a textbook that is undependable because of author mistakes. The folks at the publishing house realized—way too late—that they had 50,000 books on their hands that they simply could not sell. A hasty meeting of the MBAs was called and an executive decision made. They had to re-do the edition, but what to do with the copies they had already created? The perfect solution was obtained: they shipped them to an (unnamed) third-world country.

*T*wo colleagues were discussing a quite famous man—a mathematician, of course—his relative merits, accomplishments, prospects. There was disagreement about his ultimate place in the firmament. Finally one man said, "He is like a Euclidean point: he has position, but no mass."

$\mathcal{F}$or over 150 years, Springer-Verlag has been the grand dame of mathematics publishing. Always a family-held company, Springer produces books of the highest quality, many by the leading mathematicians of the day. We have all been educated on Springer books.

But just a few years ago Springer allowed itself to be sold to the huge Bertelsmann publishing conglomerate. This was done partly for financial reasons and partly so that Springer could take advantage of Bertelsmann's expertise in electronic publishing. After the sale was complete and the dust settled, various Bertelsmann potentates came from Germany to the United States to inspect and critique the Springer operations in America. A high-ranking corporate officer came to Emeryville, California to carry out the review of one of the West-coast Springer operations. The California folk of course wanted to treat their guest with the utmost deference, and they arranged to take him and his wife to the celebrated *Chez Panisse* Restaurant for dinner.

In case the reader is not familiar with *haut cuisine* in Northern California, *Chez Panisse* is considered by many to be the finest restaurant in the world. Begun by Alice Waters in 1971, it is the home of the concept of "California Cuisine." Alice and her staff create breathtaking meals, in a fusion of French, Mediterranean, and American cuisine, from only the freshest ingredients. It is quite an experience to dine there. The restaurant itself is a quaint wooden building with a cafe upstairs and a more formal dining room downstairs. The downstairs dinners are given in two seatings, at 6:00 PM and 9:15 PM, and the menu is fixed. You can choose any wines you wish, but the four or five courses that are served are pre-determined by the kitchen. You can view the menus a few weeks in advance on the Web so that you can choose what you like in that sense. But once you are seated in Alice Waters's dining room, the kitchen takes over.

Well, the Springer/Bertelsmann party came in and was seated downstairs with much pomp and ceremony. When the German businessman was handed the menu and was told that this was to be his dinner, he declared (in his clipped English), "When I go to a restaurant, *I* tell the kitchen what to cook for me. I do *not* let the kitchen tell me what I will eat. My wife and I will not dine tonight." And Herr and Frau sat rigidly and watched the rest of the company eat their meal. A remarkable experience for one and all, seeming to take passive/aggressive behavior to a whole new level.

It happens that Bertelsmann only owned Springer for about three years. One of the reasons the parent company ended up dumping Springer is that, around the same time that they purchased the scientific publisher, they also bought Napster (the OnLine music source). And Bertelsmann thereby lost a bundle. It needed to raise funds quickly, and Springer was one of their cash cows. So now Springer has been sold to the European venture capitalist conglomerate Cinven & Candover.

*T*oby Bailey of the University of Edinburgh fancies himself as something of a gourmet. Certainly the Guide to Good Eating that is given to all visitors in the Edinburgh Math Department is the creation of Toby himself.

A few years ago Toby was on sabbatical leave in Berkeley. One night there was a fierce thunderstorm in Berkeley, of the sort that leaves large segments of the city without power and people cowering under their dining tables for fear of their safety and wellbeing. Toby, however, is no slouch. He realized that this was just the sort of night when Chez Panisse would have cancellations. He phoned them up, got a reservation, and had a happy night out.

*F*erdinand Springer (1907–1998), the scion of Springer-Verlag during the mid-twentieth century, related that just before the end of World War II he was captured by the Russians. The Russian army major who interviewed Springer asked him who he was. "I am a scientific publisher. I publish about one hundred journals." The major asked him to write down the titles of these journals. When Springer reached title number 90, the major said, "That is enough; I have published papers in this journal and in this one." Evidently the major was a geneticist; in fact he had evidently been an editor of one of Springer's journals. Springer had refused, despite a request from the Nazi government, to remove the editor's name. After a few days the Russians let Ferdinand Springer go; but they advised him to stay with them, lest he be captured again by a less scientific Russian unit.

*I*n fact there is a more wistful version of the story about why Bertelsmann sold Springer after only a few years.

A highly placed executive of mathematical publisher Springer-Verlag explained how the breast enhancement surgery of Britney Spears led to the sale of Springer and the demise of Kluwer.

As has been previously noted, Bertelsmann is a huge publishing conglomerate, heavily invested in music. In particular Bertelsmann is heavily invested in Britney Spears. Now Britney started singing when she was quite young. When she approached the biological brink of womanhood, she essayed to change her image. This entailed breast enhancement and an attitude modification. She became sassy and irreverant. In particular, when performing on stage in Mexico she "flipped off" her audience. The reaction was immediate and negative and her record sales commenced to plummet.

Now Thomas Mittelhof, the CEO of Bertelsmann at the time, had a "put option" as part of his compensation package with the publisher. When he saw Spears's fortunes take a negative turn, he panicked and invoked the put option. Thus Bertelsmann suddenly had to raise a substantial amount of money. Springer was one of their cash cows, so the decision was made to sell Springer. The rest is the stuff of legend.

*S*aunders Mac Lane tells of teaching at Harvard in the 1930s. It was a glorious time, as among the faculty were George David Birkhoff, Magnus Hestenes, Lars Ahlfors, Julian Lowell Coolidge, William Caspar Graustein, E. V. Huntington, Marston Morse, Joseph Walsh, Marshall Stone, and David Widder. He relates that Morse, as always, was particularly enthralled with his own ideas. Mac Lane recalls "standing about street corners listening to him speak about his newest interests, such as the calculus of variations in the large. Evidently William Fogg Osgood (1864–1943) married Morse's ex-wife and left town with her. In fact it seems that they went together to China!

Morse suffered both personal and public angst over the event. The following ditty records some of the sentiment of the time:

> Here's to Marston, Mickey Morse
> A man experienced in divorce.
> His opinion of himself, we charge
> Like nose and book is in the large.

*A*lfred Tarski visited the Soviet Union in 1966 in connection with the International Congress of Mathematicians that was held in Moscow. A native of Poland, Tarski was also proud of his fluency in the Russian language. He told with relish the story of the dinner that was booked for him at the hotel restaurant in Novosibirsk. Tarski arrived punctually and seated himself. Eventually (service in the restaurant was typically *very* slow) a waiter came over and Tarski proceeded to order his dinner in Russian. The waiter stopped him and asked him to leave, declaring that this table was reserved for a famous *American* professor.

*J*ohn Horton Conway once offered a substantial sum of money—some say it was as much as $50,000—for the solution of a problem he had been thinking about (unsuccessfully) for a long time. It was the sort of problem at which Conway was *the* ranking expert. If he couldn't solve it, then nobody could. So he felt quite safe in offering this bounty. Conway was no doubt inspired by Paul Erdős, who had been doing this sort of thing for decades. People liked to ask Paul (who had $30,000 in outstanding offers for problem solutions), "What would you do if all your problems were solved? How could you pay?" Erdős would reply, "What would the world's strongest bank do if all its depositors came in on the same day and demanded their money? Of course the bank would go broke. But that is much more likely than that all my problems will be solved."

Hah! About three days after Conway posed his problem, Jeff Lagarias of AT&T Bell Labs heard it, and he produced a swift and elegant solution. Conway was dumbfounded, and he did *not* have the $50,000. He struck a deal with Jeff: He would give Jeff a check for $50,000 if Jeff would promise never to cash it. And Lagarias settled for a cash bounty of a much smaller denomination.

*A*t Harvard one day there was a group of graduate students and Peirce postdocs shooting the breeze together. One graduate student was complaining about the algebraic geometry course, given by one of the Harvard luminaries, that they were all attending. Said he, "I thought that algebraic geometry was about the zero sets of polynomials. But everything in this course is so abstract. It all seems to be a bunch of algebraic pie-in-the-sky. I can't tell what anything is. Can't someone give me a concrete example?" "Sure," replied one of the Peirces. "Let X be a pre-scheme ...."

*A*rmand Borel hosted a fancy dinner at a restaurant in Zurich in celebration of his sixtieth birthday. Armand Borel, Bill Casselman (1941–), Mark Goresky (1950–), and Robert MacPherson were present. Near the end of the evening, he pulled out an envelope with a new credit card. He made a great ceremony of removing the cognate old credit card from his wallet, breaking it up into many pieces, and putting the pieces in a pile by his empty wine glass. At this point Mark Goresky said, "Why, it appears that you have broken the wrong card." Borel gave him a defiant stare, and Goresky stared back. Goresky went on to say, "You don't seem to be very worried." "Well," said Borel, "In the first place, I checked it three times before breaking it up. And in the second place, and even more important, I am becoming familiar with your peculiar sense of humor." There was an awkward moment of silence, and then the entire table burst into laughter.

$\mathcal{A}$ndré Weil always had an easy confidence, and a high opinion of his own mathematical abilities. In his autobiography *The Apprenticeship of a Mathematician* [WEI], he is careful not to be too critical of others. But he allows such words to be slipped into the mouths of objective observers. In one instance he indicates that he asked Francesco Severi (1879–1961) what he thought of Lefschetz's work. "E bravo," said Severi [meaning "very nice"]. "But he is no Poincaré."

$\mathcal{O}$nce, some years ago, Armand Borel was giving a lecture at the Institute for Advanced Study. Of course the Institute was Borel's turf, and he addressed his audience with an easy confidence. But André Weil was in that audience, and he was no wilting flower. At one point in the lecture Borel made a mistake—as we all will do from time to time—and Weil attacked him mercilessly. Weil stood up and railed for quite some time at the hapless Armand Borel. When the great man finished, he sat down. Borel said, "So what am I supposed to do now? Commit suicide?"

$\mathcal{G}$eorg Cantor (1845–1918) was a tormented individual. His theory of sets, particularly his theory of cardinal numbers, was not universally embraced. He suffered anti-Semitic attacks, and he spent considerable time in sanatoria endeavoring to recover from bouts of depression. In 1885 Cantor was to have an article appear in the venerable *Acta Mathematica*. While it was being type-set, Cantor received a letter from the journal's founder and editor, Gösta Mittag-Leffler. Mittag-Leffler begged Cantor to withdraw his paper:

I am convinced that the publication of
your new work, before you have been able
to explain new positive results, will greatly damage your reputation among mathematicians. I know very well that basically
this is all the same to you. But if your theory is once discredited in this way, it will
be a long time before it will again command the attention of the mathematical
world. It may well be that you and your
theory will never be given the justice you
deserve in your lifetime. Then the theory
will be rediscovered in a hundred years or
so by someone else, and then it will sub-

Georg Cantor

sequently be found out that you already had it all. Then, at least, you
will be given justice. But in this way [by publishing the article], you
will exercise no significant influence, which you naturally desire as
does everyone who carries out scientific research.

In the end, Cantor did publish his article. Indeed, the work

Über verschiedene Theoreme aus der Theorie der Punctmengen in
einem *n*-fach ausgedehnten stetigen Raume G_n, Zweite Mittheilung[*]
by Georg Cantor

appears in *Acta Mathematica* 7(1885), 105–124.

*S*tefan Bergman's wife Adele doted on him. At the beach, she made sure he
was properly shaded and coated with oil. At math conferences, she followed him from room to room and catered to his every need.

It seems, however, that Stefan did not necessarily reciprocate this passion. He was stingy by nature, and his wife suffered from this fastidiousness
as much as anyone. Stefan never bought his wife a washing machine, and
she spent her whole life washing clothes on a washboard in a zinc tub.
When he died, he gave the bulk of his estate to the American Mathematical
Society to set up the Bergman Fund (which administers the Bergman Prize).

---

[*]Loosely translated, this title is *On various theorems concerning the theory of point sets in an
n-fold extended continuous space G_n, second part.*

$\mathcal{M}$ath editor Jeremy Hayhurst (1956–) tells of chatting with a rather self-confident new calculus author. This was about 25 years ago, and the man was particularly pleased with his new book because *computers* had been used to check everything—all the theorems, all the proofs, all the examples, all the spelling, all the exercises, all the solutions. This book was perfect. Nobody could fault this book for accuracy. Jeremy was duly impressed, and he fondled the sample copy with due esteem and admiration. At the appropriate moment he cracked the spine and opened to the title page. There he saw, boldly displayed in 48 point type, the title

CLACULUS

$\mathcal{L}$arry Gruman (1942–), who now works at the Universitè Paul Sabatier in Toulouse, France, grew up in Minnesota. He likes to remind us that he went to high school with the two most famous living Minnesotans: Bob Dylan and Garrison Keillor.

$\mathcal{A}$ndré Weil grew up mathematically during the depression. Jobs were hard to come by, even for such an enormous talent as Weil. In many years he was supported by grants. One of these was a CNRS grant from the French government. Such grants are quite remarkable (even today), for they have no duties attached. The recipient is paid to do whatever he/she desires. So every three months Weil would receive a government check that he would cash in Paris on the Rue de Rivoli. Weil's sister, the moral and political philosopher, author, and mystic Simone Weil (1909–1943), was quite bemused by all this. When people would ask her what her brother did, she would reply, "He does research." When asked in what, she would go on to say, "How to get money from the government."

$\mathcal{A}$fter André Weil landed in the United States in the 1940s, and after he floundered around for a while on a few meager grants, he landed a job teaching at a (at that time) not-very-distinguished university in the East. This was not a happy time for Weil, and he found the work frustrating. When he later sent his book manuscript on algebraic geometry to the American Mathematical Society for publication, the only change that they requested was the removal of a sentence in the Preface that said, "[I owe an enduring debt of gratitude to ...] and to those few who strove to liberate me from the distasteful and humiliating duties of a job which my position as a refugee in the United States had compelled me to accept."

Weil describes his teaching service at this venerable institution in words that sound all too familiar today:

> The institution to which I belonged (a word that all too accurately describes my relationship with my employer) was graced with the noble title of "university"; but in fact, it was only a second-rate engineering school attached to [XYZ Corporation]. The only thing expected of me and my colleagues—who were totally ignorant as far as mathematics went—was to serve up predigested formulae from stupid textbooks and to keep the cogs of this diploma factory turning smoothly. Sometimes, forgetting where I was, I would get carried away and launch into a proof. Afterwards, according to the well-established ritual, I would always ask, "Are there any questions?" Just as predictably would come the question: "Is that going to be on the exam?" My answer was ready: "You should know it, but it's not very important." Everybody was happy.

Some of Weil's classes at this Eastern institution were taught under the auspices of the Army Science Training Program. Students would show up in class dressed in uniform, marshaled by a non-commissioned officer. Sometimes, in order to maintain order in the classroom, Weil found it necessary to entrust the non-commissioned officer to issue a command ordering the students to attention. One day, one of the young men said, "I don't understand what x is." Weil's comment was, "The question was far more profound than he suspected, but I did not attempt to explain why."

At a moment of near despair, Weil wrote to Hermann Weyl and said, "Prostitution consists in diverting something of high value to base uses for mercenary reasons; this is what I have been doing these two years."

John Erik Fornæss (1946–) earned tenure at Princeton University in 1979, just five years after receiving the PhD. To celebrate this accomplishment, a conference was held at Princeton. Of course it was in the area of John's specialty: several complex variables.

The banquet for the conference was held at the Princeton University Faculty Club; the food and wine were, quite frankly, fairly dreadful. At the end of the meal, R. Narasimhan (1930–) was asked to stand up and deliver a toast. He rose and said, "We're out of wine. Let's go home."

*I*n the first volume of *Mathematical Apocrypha*, we told of the eminent mathematician, not possessed of a PhD, who was asked why this was the case. His reply was "Who would examine me?" In fact this story is told of a number of eminent mathematicians from the first half of the twentieth century. Richard Courant claims that indeed Garrett Birkhoff (1911–1996) of Harvard uttered these words. He goes on to say that the description also fits Gauss.

On the other hand, Gauss seems to have written a thesis at the University of Helmstadt in 1799. Jerry Alexanderson owns a copy of it. The work contains Gauss's proof of the Fundamental Theorem of Algebra. It has yet to be determined whether this document is comparable to our modern notion of a PhD thesis.

*N*orbert Wiener's father was a strong and commanding figure, and an important role model in Norbert's life. A Professor of Languages at Harvard, he was an accomplished and imposing man. Norbert was subjected to strict and demanding training (at home) under his father Leo's tutelage. According to P. R. Masani (1919–1999), certain prejudices and temperamental weaknesses of the parents affected Wiener rather drastically in his youth. Of course there were the adverse effects of his father Leo's extremely harsh training. But also Norbert was led to believe, at least until puberty, that he was a gentile. The sudden revelation at age fifteen (in the year 1911) that this was a lie was shattering. In Norbert's own words:

> The wounds inflicted by the truth are likely to be clean cuts which heal easily, but the bludgeoned wounds of a lie draw and fester.

The "black year of my life" was his description of 1911.

Norbert Wiener

$\mathcal{M}$any mathematicians like to show off, or be the center of attention, or both. Norbert Wiener often *demanded* attention. He was short, rotund, and had a neatly pointed beard that he wore in such a fashion that it pointed up in the air. It was said that he wore his bifocals upside down. He worried that his students called him "Wienie" (but in fact they called him "Norbie"). When he attended a lecture, Wiener would habitually walk in late, go down to the front row, take out a magazine, and read it ostentatiously. Then he would sleep (blatantly) and wake up at the end of the lecture to ask a point-ed question, or sometimes to offer a little mini-lecture of his own.

Wiener had a game of asking other mathematicians, or groups of math-ematicians, to make a list of the ten finest American mathematicians. At one math meeting at Duke in the late 1930s, a number of people concocted a response to this ploy. They would run briskly through a list of nine out-standing mathematicians, omitting Wiener's name, and then look thought-ful as they puzzled over the tenth name. They continued this charade until Wiener's squirming became unbearable. This sounds a bit cruel, but wit-nesses say that they suspect that Wiener knew what was going on and enjoyed the attention.

$\mathcal{H}$ans Freudenthal (1905–1990) gives the following "first glimpse" of Norbert Wiener:

> In appearance and behavior, Norbert Wiener was a baroque figure, short, rotund, and myopic, combining these and many qualities in extreme degree. His conversation was a curious mixture of pompos-ity and wantonness. He was a poor listener. His self-praise was play-ful, convincing, and never offensive. He spoke many languages but was not easy to understand in any of them.

$\mathcal{P}$. R. Masani, a great admirer and scribe of Wiener, quotes G. de Santayana (1902–1974) in describing Norbert Wiener, "In his reactions he was a child, in his judgments a philosopher." He goes on to say that, with Wiener (much as with Beethoven), all traces of immaturity and eccentricity vanished when he picked up his scholarly pen. Masani studied all of the 250+ papers of Wiener and claims that he found only three in which "Wiener-noise" damps out the Wiener message. He has also studied Wiener's voluminous corre-spondence with over 1000 individuals. These range from a prisoner in the federal penitentiary at Attica to leaders of industry and labor, and of course numerous scholars around the world. Masani encountered only one (a letter to Dr. Frank Jewett (1879–1949), in September of 1941, in which Wiener tenders his resignation from the National Academy of Sciences) that was intellectually confusing.

$\mathcal{U}$p until 1960, Harry Carver (1890–1977) was the dominant figure in statistics at the University of Michigan. He had a quick mind and a warm and sympathetic manner. He worked very closely with his students, and his students were fond of him. He had many eccentricities; for example, his diet seemed to consist primarily of crackers and milk. He frequently would offer to buy his entire class dinner if it could beat him at one of five indoor sports—card games or billiards or pool—or at one of five outdoor sports such as running or putting the shot. He never lost.

I am somewhat ashamed to admit that I once had a student in my office complaining about his grade. I offered to play him a game of pingpong for the grade: if he won he got an "A," otherwise he was stuck with a "B." The student was so intimidated that he declined.

$\mathcal{G}$. D. Birkhoff (1884–1944) was an imperious teacher, who often lectured with stage props both large and small. In the summer of 1929 he gave a course at Columbia University on "Mathematical Elements of Art." The last lecture of the class was on music, and Birkhoff insisted that the lesson be delivered in a special room (202 Hamilton Hall) which contained a lovely, ebony Steinway piano. Birkhoff lectured for the entire period without ever touching, or even referring to, the piano.

$\mathcal{O}$n another occasion, Birkhoff gave one of the inaugural lectures for the opening of the Institute for Advanced Study. His lecture was a tour de force of technical mathematics, and the blackboard rapidly filled with exotic symbols. He suddenly stopped short, looked up and down, and said with great surprise, "But there is no colored chalk here." The Princeton hosts expressed great consternation and concern, and a young local professor was rapidly dispatched to go find some colored chalk. While the young fellow raced away, Birkhoff continued with his lecture. Eventually the expedient returned with a beautiful display box of every possible color of chalk and presented it to the august lecturer. Birkhoff looked at him over his spectacles and said, "That's all right. I don't plan to use it," and he went on with his presentation.

$\mathcal{L}$ars Ahlfors fondly recalls his education at the feet of Ernst Lindelöf (1870–1946) and Rolf Nevanlinna (1895–1980). At the time, Lindelöf had retired from research, but he remained a devoted and excellent teacher.

Advanced students were expected to come to Lindelöf's home on Saturdays at 8:00 AM to be either scolded or praised for the quality of their written assignments in the past week. Lindelöf had a considerable mathematics library, with books in many languages, and he lent the books freely to his students. He never asked the borrower whether he/she could read the language in question. It was just expected that they would do so. Lindelöf particularly encouraged Ahlfors to read the works of Schwarz and Cantor, but he strenuously warned Ahlfors *not* to become a logician. Ahlfors recounts that he was always most grateful for that advice. At the time, Riemann was considered too difficult, and Lindelöf never quite approved of the Lebesgue integral.

$\mathcal{P}$aul Rosenbloom (1920–) did his graduate studies at Stanford in the 1940s. He tells this story of his experience:

> In January 1943, I received an offer of an instructorship at Brown, providing that I could start in February. Gàbor Szegő (1895–1985) arranged for me to take my final orals immediately, even though my thesis wasn't written yet. I was asked to outline my main results thus far, and then the committee probed me with general questions. Pòlya began by asking me to give the definition of Gaussian curvature in terms of the area of a spherical map by the normals. I protested that I had never studied it, but he insisted that I try to work it out at the blackboard. He said, "It is not forbidden to learn something from an examination."

$\mathcal{J}$ulian Coolidge's rather lofty views of Harvard education ran somewhat counter to the philosophy of President Eliot (1834–1926) of the university.

In the year 1900, for instance, Harvard offered a number of very popular "gut" courses. After 30 years of President Eliot's unstructured "free elective" system, it was possible to get a Bachelors Degree from Harvard in just three years with very little effort indeed. In addition, it was to be noted that athletic excellence was widely admired by the student body; scholarly excellence was not. Someone who worked hard at his studies might be called (by his peers) a "greasy grind." A social rift had developed between "the men who studied and the men who played."

$\mathcal{S}$aunders MacLane described von Neumann's passion for precision, as manifested in the way that he learned Hilbert space theory. According to MacLane, "Two of von Neumann's papers on this topic had been accepted in the *Mathematische Annalen*, a journal of Springer Verlag. Marshall Stone had seen the manuscripts, and urged von Neumann to observe that his treatment of linear operators T on a Hilbert space could be much more effective if he were to use the notion of an adjoint T^* to the linear transformation T—one for which the now familiar equation

$$\langle Ta, b \rangle = \langle a, T^* b \rangle$$

would hold for all suitable a and b. Von Neumann saw the point immediately, as was his wont, and wished to withdraw the papers before publication. They were already set up in type; Springer finally agreed to cancel them on the condition that von Neumann write for them a book on the subject—which he soon did …"

$\mathcal{T}$oday, at research Math Departments, the standard teaching load for tenure-track faculty is two courses per semester (or less!). The understanding is that a mathematician must have considerable blocks of time to concentrate on his scholarly work and the training of graduate students. But this has not always been the case. There is general agreement that in the late 1950s and early 1960s—during the Sputnik era—there was great competition for top mathematicians and lower teaching loads were among the perks developed to attract the top people. Prior to that time, everyone (even at the best institutions) taught a lot. Ivan Niven tells of how things were seventy years ago:

> Teaching loads were higher in the thirties. In schools with PhD programs, three courses was a common load for younger faculty mem-

bers, except at a very few major private institutions with slightly lighter loads. But four courses was the common load at institutions with only a Master's degree program or no graduate work at all. Emeritus Professor M. Wiles Keller (1905–1993) writes that when he went to Purdue University in 1936, "most of the staff taught 18 hours per week." He added that loads of 15 hours were possible, presumably for a few more scholarly professors. This differential in teaching loads for scholars was not uncommon: Ralph P. Boas reports that at Duke University he was given only 3 courses "as an incentive to research," where 4 courses was the nominal teaching load. Similarly, Abraham H. Taub (1911–1999) writes that he went to the University of Washington as an instructor in 1936 with a teaching load of 13 hours a week, where the normal teaching load was 15 hours a week. He was given a "research" allowance.

According to the AMS Survey ... headed by A. A. Albert (1905–1972), teaching loads in the mid-fifties were still around 10 or 11 hours per week for younger faculty members in the major state schools with PhD programs.

$\mathcal{H}$ans Rademacher (1892–1969), a distinguished professor in analysis and analytic number theory, had been a full professor of mathematics at the University of Breslau, Germany from 1925 to 1933. He was dismissed by the Nazis. He was one of the lucky ones, however, for he landed at the University of Pennsylvania. But he only had the rank of Assistant Professor. He kept hoping for promotion to a rank that was in accord with his accomplishments, but (according to Emil Grosswald), "In those years, the length of faithful service to the institution ... and not professional excellence, was the main criterion for promotions—a fact that was forcefully explained to the somewhat surprised assistant professor by a most self-assured dean."

Ivan Niven goes on to tell us that, "The University promoted Hans Rademacher from an assistant professor to a full professor in 1939, to meet an outside offer from a comparable university. There was a slight hitch when the offer came in, since the considerable salary raise needed to keep Rademacher amounted to virtually the entire dollar amount for raises for the whole department. John R. Kline, the head of the graduate program in mathematics, pressed the University administration very hard that it would be a serious mistake to lose such an outstanding scholar who was at the same time a superb teacher. Kline succeeded; funds were found and Rademacher stayed on."

"To be a mathematician, one must love mathematics more than family, religion or any other interest."

—Paul R. Halmos (1916–)

*S*ome feel that the University of Chicago Math Department went into a period of second-rate work in the 1920s and 1930s. Saunders Mac Lane has suggested that this was because too many of the retiring eminent faculty were replaced by their own PhD students. Evidently this slippage was not universal across campus. As W. L. Duren, Jr. puts it:

> It was no ebb cycle for the University of Chicago as a whole in the twenties. There was intellectual excitement in many places in the university. I attended the physics colloquia where the great innovators of the day came to talk. With Mr. Bliss's (1876–1951) grudging consent, I took Arthur Holly Compton's (1892–1962) course in X-rays. He already had the Nobel Prize for his work on the phenomena of X-rays colliding with electrons. Yet he seemed so naively simple minded to me, far less expert and mentally profound than other physicists in the department. Somewhere in here Einstein came for a brief visit. He permitted himself to be escorted by the physics graduate students for a tour of their experiments. To one he offered a suggestion. The brash young man explained immediately why it could not work. Einstein shook his head sadly. "My ideas are never good," he said.

*D*avid Widder tells of when, in 1936, Hardy was a guest at Harvard. There he presented a lecture on number theory. The hostess at the house where Widder resided allowed the young man to invite Hardy to be a guest at dinner. In the course of the evening the eminent mathematician had occasion to use his fountain pen, and he left a stain on the hostess's gold-colored sofa. The gracious lady said that she would treasure that spot, reminding her of the great man's visit.

*J*ohn Milnor first came to Princeton in 1948 at the age of seventeen. By his own telling, he was socially inept and so felt comfortable and welcome in the friendly and warm atmosphere of the Fine Hall Mathematics Department. He speaks particularly warmly of Ralph Fox, who was both Milnor's senior thesis advisor and his PhD thesis advisor. Of Fox he says, "Fox was a wonderful teacher in a strange way. I believe that his lectures were often somewhat confused: he didn't always get things exactly right. The effect was that the class had to help him. (I am afraid that I have often unintentionally followed his example.) However we learned a great deal in this way.... Let me mention another teacher:

Emil Artin, who was an extremely impressive and charismatic figure. His style was the exact opposite [of Fox]: each lecture was a totally polished work of art. There was not a misplaced epsilon, and everything was exactly efficient and perfectly done. In theory, this should be much more conducive to learning but in practice I am not sure that it was. For example, I attended a course in algebraic number theory from Artin which was extremely elegant, although perhaps too advanced for me. However, it wasn't until a few years later that I learned what an algebraic number was. The course was so streamlined that algebraic numbers were never actually mentioned."

*O*f course the great father figure and provider in the Princeton math department in Milnor's young days was Solomon Lefschetz. Milnor says, "If Ralph Fox was my mathematical father, then Solomon Lefschetz was my mathematical grandfather. Lefschetz was an amazing figure—I don't know how else to put it. He had to work with very clumsy artificial hands, which could hold a piece of chalk but not much else. Don Spencer tells us that his wife called him Sol, but I never heard anyone call him by his first name. He looks rather imposing…, and he was imposing in person. It was not that he intended to be: he was very friendly and helpful. I remember that he would fix me with this gaze and give me all sorts of wise and useful advice, but I didn't know quite how to take it. In fact I am afraid that sometimes I had to work hard to keep a straight face. I admired him greatly: I think he spoke all known languages, at least all languages I'd ever heard of, although as far as I could tell he had a strong accent in every one of them. And he seemed to know every part of mathematics—he had worked in algebraic geometry, topology, and the qualitative theory of differential equations (what we now call dynamics). He wrote books on many different subjects, and I am afraid that he was confusing on every subject, but his work was nevertheless useful and influential."

*N*icholas Th. Varopoulos of the University of Paris used to tell of a great theorem he proved one day. He had been working on it a long time, and was elated to have finally cracked it. In fact he was so happy that he dropped everything he was doing and went to the Swiss Alps to do some skiing and to celebrate. Late one afternoon, he was sitting on a balcony overlooking the Alps, eating a plate of spaghetti, and pondering the beauty of this wonderful new result he had discovered. He suddenly realized that there was a mistake. On further thought, he realized that the error was fundamental. It could not be fixed. Varopoulos got up, put his skis back on his car, and left. He says that, for all he knows, the plate of spaghetti is still sitting there.

*I*n the old days at Princeton, when the Institute for Advanced Study was just starting up and the new members were waiting for Fuld Hall to be built out at the new Institute grounds, the Institute scions shared their quarters with the mathematicians in old Fine Hall on the Princeton University Campus. One of these luminaries was Albert Einstein. Of course Einstein fancied himself to be a violinist. And he used to whip out his violin late in the afternoon each day and saw away—much to the consternation of the others in the building who were trying to work.

One day Einstein was at home rehearsing a Haydn string quartet with some friends. When the great man failed for the fourth time to get his entry in the second movement, the cellist looked up and said, "The problem with you, Albert, is that you simply can't count."

*W*hen Charlie Fefferman, Fields Medalist, was at the height of his celebrity, there was a feature article about him in *People Magazine*.[*] The article reveals many of Charlie's closely kept secrets: that he throws away 90% of his work, and that he does much of his best work lying on his back on the sofa with his eyes closed. A number of mathematicians have remarked that they have tried this last technique, but their spouses always accuse them of sleeping.

---

[*] *People Magazine*, Oct. 15, 1979, pp. 73–74.

$\mathcal{A}$t Washington University in St. Louis we like to tell our calculus students (who are a diverse lot, coming from across the country and around the world) that Missouri was the home of America's greatest poet, greatest novelist, greatest playwright, and greatest mathematician. This assertion usually garners looks of bewilderment, and the reader may find himself/herself unable to identify some of these individuals. They are:

Greatest Poet: T. S. Eliot (1888–1965)

Greatest Novelist: Mark Twain (1835–1910)

Greatest Playwright: Eugene O'Neill (1888–1953)

Greatest Mathematician: Norbert Wiener (1894–1964)

In fact Washington University was founded in 1853 by T. S. Eliot's grandfather William Greenleaf Eliot (1811–1887); the institution was originally known as the Eliot Seminary.

$\mathcal{T}$. S. Eliot's widow, Valerie Eliot, once wrote this letter to the London *Times*:

My husband, T. S. Eliot, loved to recount how late one evening he stopped a taxi. As he got in the driver said, "You're T. S. Eliot." When asked how he knew, he replied, "Ah, I've got an eye for a celebrity. Only the other evening I picked up Bertrand Russell, and I said to him, `Well, Lord Russell, what's it all about?' and, do you know, he couldn't tell me."

$\mathcal{I}$n the early 1950s, Steve Hu (1914–) at UCLA was a very active and energetic young mathematician. He wrote, on average, a paper per month. And he submitted them all to the venerable *Annals of Mathematics*. Hu was usually favored with one of three replies: **(i)** the paper was accepted (obviously the most desirable outcome), **(ii)** the paper was rejected (a somewhat less attractive turn of events), or **(iii)** the editor asked for revisions. Now Hu was quite busy cranking out a paper per month, so he had no time nor patience for outcomes **(ii)** or **(iii)**. In those eventualities, he would send the paper to *Mathematicae Portugalae*—a journal which in those days had the reputation of publishing just about anything. Hu was a fine mathematician, and quite a few of his works were published in the *Annals*. Many of the others enjoyed the fate just described. So if one examines the publication record of

Steve Hu (as is easily done with `MathSciNet`) one sees a long list of papers in the *Annals* alternating with papers in *Mathematicae Portugalae*.

Jn 1939, Thomas Doyle (1908–2002), Ralph Fox, and Robert E. Greenwood (the source of this story) of the Princeton Math Department gave a beer party. This was customary for people receiving their doctorates. The party was at the Nassau Tavern. Dean Eisenhart (1876–1965, also a mathematician, after whom the Eisenhart Arch at the Graduate College is now named) attended and asked for a glass of ginger ale. Two notorious tee-totalers, John Tukey (1915–2000) and Robert Eddy (1914–1983, a chemist), also gave a party for the finishing doctorates that year. Their party was a milk party. They had big milk jars, and they poured out milk for the attendees.

Raoul Bott (1923–) left the University of Michigan in the early 1960s to move to Harvard. He recalls having lunch in the Harvard faculty club and seeing Henry Kissinger (1923–) always working the room, pressing the flesh. One day one of his colleagues leaned over, pointed to Kissinger, and said, "You see. That is our very own Dr. Strangelove."

Today, after being a leading luminary in the Harvard Mathematics Department for four decades, Raoul Bott is retired. He resides in—and is indeed the mayor of—Menemsha Beach on Cape Cod. One of the chief features of this town is a nude beach at which the dress code is strictly enforced (those wearing clothing are banished). Bott rules the roost in the official attire.

René Descartes was a short (5′0″ dripping wet), irascible Frenchman who was also one of our greatest philosophers and mathematicians. He thought very highly of himself and his abilities, and he had little patience along with a blazing temper. Descartes lay abed each day, naked, until 11:00 AM—meditating and formulating his ideas.

Descartes gave up mathematics when he was still a young man because, he said, he'd gone as far in mathematics as a human being could go. He read many romance novels and novels of chivalry; Descartes had an active fantasy life. He had a particular fetish for cross-eyed women.

Among his more unusual beliefs was the contention that animals were "senseless machines." Even so, Descartes had a pet dog of which he was very fond.

Descartes used to play cards and gamble with his friend Blaise Pascal (1623–1662). It is said that Descartes made a lot of money thereby.

*O*ne day John von Neumann was scheduled to give a lecture. He marched forcefully into the room, went straight to the blackboard, and began by saying, "The goys have proved the following theorem …"

Utter Seriousness

*W*hen Henri Poincaré took the entrance exam to the Collège de France, he was asked the following question:

Determine the next term of the sequence

$$2, 3, 7, 9,\ldots$$

Like any good mathematician, Poincaré found the question frustrating and annoying. He said, "There is no unique answer. There is not sufficient information." Poincaré answered several other exam questions in a like manner.

The examiners were not amused, and they failed Henri Poincaré on his first attempt to gain admission to the illustrious College.

*O*ne of the great modern exponents of finite mathematics, combinatorics, group theory, classical geometry, and many other parts of "hands-on" mathematics is John Horton Conway of Princeton. One day a student of John's presented him with this sequence:

```
1 3
1 1 1 3
3 1 1 3
1 3 2 1 1 3
1 1 1 3 1 2 2 1 1 3
3 1 1 3 1 1 2 2 2 1 1 3
1 3 2 1 1 3 2 1 3 2 2 1 1 3
1 1 1 3 1 2 2 1 1 3 1 2 1 1 1 3 2 2 2 1 1 3
```

and so forth. The challenge was to find the rule that generates the sequence. What is the next element?

Now if anyone can solve a problem like this it is John Horton Conway. But he couldn't do it. After a couple of weeks of effort he finally had to throw up his hands and ask for the answer.

This wonderful sequence has come to be called the "Look and Say Sequence". The rule is this: We begin with the "word" **13**. Then we read it aloud: "One 1 and one 3." This produces the new sequence **1113**. Now we read this new sequence aloud: "Three 1s and one 3." This yields the new number **3113**. Continuing in this fashion we obtain the displayed sequence of numbers.

The fecund mathematical mind will immediately ask many questions about this sequence. First, what can we say about the length of the kth string? John Conway came up with the startling answer that the size has growth asymptotically equal to $C\lambda^k$, where $\lambda = 1.3035772690\ldots$ is the largest real zero of the polynomial

$$\begin{aligned}
p(x) = {}& x^{71} - x^{69} - 2x^{68} - x^{67} + 2x^{66} + 2x^{65} + x^{64} - x^{63} - x^{62} - x^{61} \\
& - x^{60} - x^{59} + 2x^{58} + 5x^{57} + 3x^{56} - 2x^{55} - 10x^{54} - 3x^{53} - \\
& 2x^{52} + 6x^{51} + 6x^{50} + x^{49} + 9x^{48} - 3x^{47} - 7x^{46} - 8x^{45} - 8x^{44} \\
& + 10x^{43} + 6x^{42} + 8x^{41} - 5x^{40} - 12x^{39} + 7x^{38} - 7x^{37} + 7x^{36} \\
& + x^{35} - 3x^{34} + 10x^{33} + x^{32} - 6x^{31} - 2x^{30} - 10x^{29} - 3x^{28} + 2x^{27} \\
& + 9x^{26} - 3x^{25} + 14x^{24} - 8x^{23} - 7x^{21} + 9x^{20} + 3x^{19} - 4x^{18} \\
& - 10x^{17} - 7x^{16} + 12x^{15} + 7x^{14} + 2x^{13} - 12x^{12} - 4x^{11} - 2x^{10} \\
& + 5x^9 + x^7 - 7x^6 + 7x^5 - 4x^4 + 12x^3 - 6x^2 + 3x - 6.
\end{aligned}$$

It turns out that the very same constant λ applies to *all* such sequences, regardless of the starting string, with just two exceptions. These are **(a)** the empty initial string and **(b)** the string **22**. Ekhad and Zeilberger have the definitive proof of this last result.[*]

There has actually developed a theory of "atomic sequences" à la Conway. An "atom" is a piece of a sequence element whose subsequent development (in later elements of the sequence) does not interfere with nearby elements. An example of an atom is **22**. In the next generation it gives rise to **22** and so on *ad infinitum*. All Conway sequences are composed of atoms in a certain precise sense. There are 92 special atoms (named after the chemical elements hydrogen, helium, etc.) The asymptotic density of these atoms has been computed. For example, of every million atoms that occurs in nature, about 91790 of them will be of hydrogen. We

---

[*]These two mathematical stars appear elsewhere in the present book. Recall that Shalosh B. Ekhad is Zeilberger's nickname for his computer.

provide a brief table of some of the more common elements:

Element Name	String	Frequency per Million
H (Hydrogen)	22	91790
He (Helium)	3122113222122211211123222112	3237
O (Oxygen)	31221123222112	6537
S (Sulphur)	132112	19417

An authoritative source for more information about Conway's atomic theory is the URL www.btinternet.com/~se16/mhi/Part1.htm.

The number λ indicated above—now known as Conway's number—is the unique largest eigenvalue of the 92×92 transition matrix M whose (i,j)th element is the number of atoms of element j resulting from the decay of one atom of element i. There is a considerable literature on the atomic theory stemming from Conway's ideas. This entire subject area is a beautiful example of how a good mathematician can snatch victory from the jaws of defeat.

*W*riting books entails dealing with copy editors. The copy editor filters your prose with an exacting standard, checking for English usage and consistency—both locally and globally. In my first book, the copy editor made two notable recommendations. Number one, he observed that in the first chapter I made reference to a "Riemannian metric." Therefore he concluded that it was incorrect to speak of the "Riemann mapping theorem" in Chapter 10. I ought to call it the "Riemannian mapping theorem." He also took umbrage with my using the terminology "the Green's function." His suggestion was that I call it "the function of Mr. Green." I was able to overrule him on both these points.

In a more recent book on complex analysis, the copy editor had his way with the chapter on the prime number theorem. Of course the Riemann zeta function played a prominent role in that chapter. The copy editor changed every occurrence of the phrase "critical strip" to "critical ship." The mind boggles to imagine what he was thinking when he made that alteration. Fortunately we were able to change it all back.

*H*orst Tietz (1921–) recalls the trials of trying to get a mathematics education during the Hitler years in Germany. Many of the best teachers were

Jews, and they were persecuted by the government, by the university administration,
and by the students at the university. Many of
the faculty and students wore uniforms and
were quick to snap off Nazi salutes and shout
"Heil Hitler!" at the slightest provocation.
One of the quiet heroes was Erich Hecke
(1887–1947).

Hecke did not follow the strict order of
beginning his classes with the Nazi salute. He
instead entered the room and nodded his head
silently and in a friendly manner. When
groups of students would pass Hecke in the

Erich Hecke

halls and offer the Nazi salute, he would turn around with a surprised and
thoughtful look, raise his hat, bow slightly, and say, "Good morning, ladies
and gentlemen." When Hecke saw people—at the train station, for instance—
wearing the yellow Star of David, he would raise his hat respectfully. Said
Hecke, "For me the Star of David is a medal: the Ordre-pour-le-Sémite."

*J*n December of 1940, for purely political reasons, Tietz was "exmatriculat-
ed" from the university. From his point of view, he no longer had a life.
Hecke, Hans Zassenhaus (1912–1991), and B. L. van der Waerden (1903–
1996) became Tietz's heroes during this trying time. They encouraged him
to be an illegal student, and to attend lectures anyway. He had to sneak in
and out of the math building, often leaving under cover of darkness, to
avoid the Gestapo. But he succeeded for a year and a half. He was ultimately caught and then denounced. When Hecke heard that Tietz could no
longer attend his number theory lectures, he cancelled the class and refund-
ed the tuition to all the students.

Another one of Tietz's heroes was Wilhelm Maak (1912–1992), who lectured on number theory. Maak had an eccentric sense of humor and a great
inclination for fun. He was quoted as saying things like, "For a mathematician there is nothing worse than not knowing what he ought to be thinking
about." or "I wish I were two puppies and could play with each other."

Maak had a penchant for misadventures, and he enjoyed them immense-
ly. For instance, once he visited Harald Bohr's house in Copenhagen and
left behind the key to his mailbox when he departed. When he arrived

home, he could not get into his very full letter box. But he did not want to break it open. So he contacted Bohr, and Bohr sent him the key—in a letter!

*H*ecke once tried to explain to his butcher that the circle cannot be squared (using ruler and compass). The butcher ended up complaining to the Reich Minister of Culture; the conclusion of his missive asserted that, "German scientists still do not seem to have realized that for the German spirit nothing is impossible!" In later years, Hecke kept this letter framed in his office.

At one point Hecke complained to Springer-Verlag that the second volume of the classic book by Courant and Hilbert (*Methods of Mathematical Physics*) was then published in Germany; but not the first. Springer replied, "The first volume was published in 1930, the second in 1937; in 1930 Courant was a German Jew, but in 1937 he was an American citizen." Hecke's comment was, "The fact that inhumanity is coupled with so much stupidity makes one feel almost optimistic in a dangerous way."

*W*hen Stephen Hawking was a student at Oxford, he and his classmates were given a difficult set of physics problems. Two of the gang worked all night and managed to answer one and one half problems. Another young fellow got just one problem all by himself. Hawking did nothing.

Next day after breakfast, Hawking retired to his room. Three hours later he came out to go to class, and his friends began to tease him. "So, Hawking, how many problems have you got?" "I've only had time," said the budding genius, "to finish the first ten of them."

*T*his bit of insight into the mathematical existence comes from Martin Davis's Foreword to Matiyasevich's book *Hilbert's Tenth Problem*, published by MIT Press:

> My greatest insight turned out to be a thought I uttered in jest. It was known that the unsolvability would follow from the existence of a single Diophantine equation that satisfied a condition formulated by Julia Robinson. However, it seemed extraordinarily difficult to produce such an equation . In my lectures [during the 1960's] I would emphasize the important consequences that would follow from either

a proof or a disproof of the existence of such an equation. Inevitably during the question period I would be asked for my own opinion as to how matters would turn out, and I had my reply ready: "I think that Julia Robinson's hypothesis is true, and it will be proved by a clever young Russian."

Early in 1970, a telephone call from my old friend Jack Schwartz informed me that the "clever young Russian" I had predicted had actually appeared. I met Yuri a few months later at the International Congress of Mathematicians in Nice, where he was an invited speaker. I was finally able to tell him that I had been predicting his appearance for some time.

$\mathcal{T}$he University of Missouri in Columbia has a dandy math department. This fact is due in part to the awesome hiring efforts of Chairman Elias Saab. Perhaps it is also due to the fact that

1. Norbert Wiener was born on their campus;
2. Thomas Jefferson's memorial resides there.

$\mathcal{T}$he mathematician Jean van Heijenoort (1912–1986) was located at Stanford for several years. He led a remarkable and diverse life. For example, he was Leon Trotsky's (1879–1940) secretary and bodyguard in Turkey from 1932 to 1939. This is all the more remarkable because van Heijenoort was a slight and unimposing figure.

In 1986, van Heijenoort was observing the work on cleaning out Pólya's office. This was a nontrivial task, and took a couple of weeks. The estimable van Heijenoort asked whether he might go through the reprint collection, seeking items of mathematical interest; but he quickly added that he was going to Europe imminently and would sort the reprints on his return. And off he went. After he had been in Paris for a while, his wife contacted him and said that she needed him in Mexico City—would he please come? Of course van Heijenoort complied, and when he arrived in Mexico his wife shot and killed him.

$\mathcal{G}$eorge Leitmann (1923–) is a distinguished applied mathematician, now retired, from U. C. Berkeley. He has been Managing Editor of *The Journal of*

Mathematical Analysis and Applications. He is part owner of two restaurants. He is a member of a league of Italian food tasters. George has led a rich and varied life. He has traveled over 2 million miles in the past fifty years.

One of George's distinctions is that, after the scoundrel's apprehension, he was the person who debriefed the Nazi Adolf Eichmann (1906–1962). George claims that he was selected for the task because he was the only person available who could maintain his composure.

$\mathcal{A}$n erstwhile member of the University of Chicago Mathematics Department was reminiscing about some of the characters who people old Eckhart Hall. He singled out a particular topologist who was fond of bragging that U. of C. had the greatest topology group in the world. The speaker said, "Why, the University of Chicago does not even have the greatest topology group in Cook County."

$\mathcal{A}$ny of us with a bit of patience gets pestered from time to time by folks who can only be characterized as "mathematical cranks"—people with sketchy mathematical training who think that they have proved that the integers don't exist. Or some such theory. Jerry Folland of the University of Washington, a heck of a nice guy, gets perhaps more of these people on his doorstep than most of us. But he tells me that he has a trick for handling them: He *introduces* them to each other. That way they can bend each other's ears, and he can gracefully ease out of the loop.

He tells of one particular supplicant who has a theory about bases for number systems. Of course we are all familiar with base 10—the most commonly used number system in everyday discourse. Base 2 is important for computer science, and certainly base 16 has its place. But one might well ask which is the most "efficient" or "propitious" number system. The theory is that one can measure this by thinking about old-fashioned calculators that would have a column of 1s, a column of 2s, and so forth. Which number system would result in the fewest number of 1s and 2s and so forth to represent the greatest gamut of numbers? This is a calculus problem that anyone can solve, and we encourage you to do so. The answer is that "base e" is the best. Now go forth and sin no more.

$\mathcal{M}$athematician J. D. Stein (1941–) has a theory about horseracing. According to him, your best bet at the track is to bet the favorite to show (i.e., to come in first, second, or third). It seems that this position is always under-bet, and therefore is favored by the pari-mutuel system.

$\mathcal{I}$ was once visiting Yale University. I gave a talk, and afterward Peter Jones and Dick Beals took me to dinner. Jones and Beals had been colleagues at the University of Chicago before they both decided to move to Yale. In the middle of our meal, a middle-aged gentleman came to our table and introduced himself. He told them that *he* had been a tenured full Professor of Mathematics at the University of Chicago. But he woke up one morning and realized that he would never be a mover-and-shaker in the world of mathematics. He simply did not possess that *sine qua non* that true mathematical excellence required. So he quit mathematics and went into business. Now he was a tenured Professor of Business at Yale. With a big smile he told us that, "Now I'm really hell on wheels. Just what I always wanted to be."

$\mathcal{P}$eople have had a sense for some time that Spencer Bloch (1944–), distinguished professor of mathematics at the University of Chicago and member of the National Academy of Sciences, is well off. He lives in a fine residence in Hyde Park, and he seems to worry about money a bit less than the rest of us. Turns out that he comes from the family that manufactures MailPouch tobacco.

Bloch once wrote an astonishing review of the book *Étale Cohomology* by J. S. Milne [BLO]. It appeared in the *Bulletin of the AMS* in 1981. In this tract, Bloch compares the subject matter with various intimate parts of the female anatomy. He waxes quite euphoric in his comparisons, and the reader comes away bewildered, to say the least. The editor got in quite a bit of hot water for letting that one get by him.

$\mathcal{I}$saac Newton lived a rather isolated and solitary life. After he quit Cambridge and moved to London, he had regular contact with Catherine Barton, a niece who kept his London house for him.

Newton brought with him to London a large metal box in which were stored his manuscripts and papers. He was beginning a new life in the capital of England, and these were a relic of his old scholarly life. He rarely, if ever, referred to the old papers. But he kept them as an archive. He bequeathed the metal box to Catherine Barton, and she passed them on— entirely undisturbed—to future generations. The box ended up in the hands of Lord Lymington, an obese and obnoxious neo-fascist who was active in the pro-Hitler movement during World War II.

In order to raise money for his "cause," Lymington broke up Newton's archive into 329 separate lots and had them sold at Sotheby's auction gallery. He only got 9000 British pounds for the aggregate. None other than John Maynard Keynes attended the Sotheby auction and bought as much of these treasures as his meager resources would allow. Other parts of the treasure were scattered to the winds.

The good news is that, thanks to great effort on behalf of a number of good people, Newton's papers have been almost completely reassembled. The Newton archive is now available in a few public libraries and universities. The materials can be accessed for scholarly study and future learning. Among these treasures are various biographical manuscripts. Of particular interest is one by Newton's cousin Humphrey Newton describing Newton during the period that he was writing the *Principia*.

In 1989, I was the head of the organizing committee for a three-week, NSF-sponsored, American Mathematical Society Summer Research Institute. At some point, Soviet Academician Anatoli Vitushkin told me that he would be my "Russian organizer" and help me to get some-top notch young Russians to attend. Recall that this was in the bad old days, before the fall of the Soviet block. It was quite difficult for Soviet scientists to travel. I was grateful for Vitushkin's help.

And, indeed, we were able to invite four outstanding young Soviet mathematicians to our conference. At one point, however, I had to have serious dealings with the U. S. State Department. They wanted to know *why* we wanted to have Russkis at our affair, and what benefit we hoped to accrue therefrom. I had to write more than one document addressing these issues. Then one day they gave me a phone call because they still were not satisfied. I was subjected to more than an hour of questioning.

At one point my interlocutor from the State Department asked, "Will these Russians have access to any high-speed computing equipment?" Anyone who knows even a little about the state of computing in 1989 might consider this question laughable, but there you have it. "No," I said. "Well, then, what is it that you are going to be *doing* for three weeks?" I replied, "We will be talking about mathematics." End of discussion.

*T*he modern idea of the peer-reviewed scientific journal was first created by Henry Oldenburg in the year 1665. He was the founding editor of *The Philosophical Transactions of the Royal Society of London*. At the time it was a daring but much-needed invention that supplanted a semi-secret informal method of scientific communication that was both counterproductive and unreliable. Today journals are part of the fabric of our professional life. Most scientific research is published in journals of some sort.

*K*urt Gödel's favorite movie was *Snow White and the Seven Dwarfs*. Gödel said, "Only fables present the world as it should be and as if it had meaning." He never convinced his friend Albert Einstein to see the movie. Likewise, Einstein never convinced Gödel to sample Beethoven or Mozart. It is not known which was Gödel's favorite dwarf.

Kurt Gödel

One day Gödel's wife put a pink flamingo in their front yard. Gödel pronounced it to be *furchtbar herzig* (awfully charming).

*T*here is new information about the relationship between Einstein and Gödel. Both Einstein and Gödel were, in a sense, refugees from Europe. They both felt a bit out of place in Princeton, but Einstein adapted perhaps more readily than Gödel. The great physicist is quoted as saying, "Princeton is a wonderful piece of earth, and at the same time an exceedingly amusing ceremonial backwater of tiny spindle-shanked demigods."

Einstein and Gödel could hardly have been more different. The former was famous for dressing in rumpled clothing that was just one step above the ragpile; the latter typically sported a white linen suit and matching fedora. Einstein was a gregarious character who loved to have fun and tell jokes.

Gödel was solemn, solitary, pessimistic, and ultimately paranoid. Einstein enjoyed heavy German cooking, and he freely indulged his appetite for winerschnitzel and sachertorte. Gödel subsisted on a strange combination of butter, baby food, and laxatives. Gödel kept a careful diary of his food intake and daily health status; he used a strange and unusual form of German shorthand that is now extinct.

Einstein enjoyed the company of other people. He was a lively presence at parties. Gödel believed in ghosts; he had a morbid dread of being poisoned by refrigerator gases; he refused to go out when certain distinguished mathematicians were in town, apparently because he feared that they would try to kill him. Gödel said, "Every chaos is a wrong appearance."

But colleague Freeman Dyson observed that "Gödel was ... the only one of our colleagues who walked and talked on equal terms with Einstein." Gödel was undaunted by Einstein's reputation and was not afraid to challenge his ideas. Einstein found this refreshing. Of course they conversed in German, and both found that to be cathartic.

Both men were generally perceived to be on a plane higher than the rest of us. Gödel biographer Rebecca Goldstein relates that "I once found the philosopher Richard Rorty standing in a bit of a daze in Davidson's food market. He told me in hushed tones that he'd just seen Gödel in the frozen food aisle." As a result of their rarification, both Einstein and Gödel were lonely. United by a shared sense of intellectual isolation, they found comfort in each other's company. One member of the Institute community noted that the two men did not seem to want to speak to anyone else; they only wanted to converse with each other. Gödel was born in 1906, the year after the publication of Einstein's four great papers that shook the world of physics. So Einstein was old enough to be Gödel's father. But they were close personal friends.

We recounted in *Mathematical Apocrypha* how Einstein and Oskar Morgenstern accompanied Gödel to Trenton to apply for his U.S. citizenship. Gödel was terribly nervous, in part because of certain logical inconsistencies that he thought he had discovered in the Constitution. Einstein told jokes during the entire trip in an effort to calm the logician down. Morgenstern kept quiet and drove.

One of the favorite topics of conversation for these two geniuses was politics. Einstein was an ardent supporter of Adlai Stevenson, and he was considerably vexed by the fact that Gödel voted for Dwight Eisenhower.

$\mathcal{C}$ertainly relativity theory was a frequent and passionate topic of conversation for Einstein and Gödel. Gödel was particularly fascinated by the concept of "time." Whereas Einstein believed that time was a relative concept, Gödel believed that time did not exist. He rejected the purely verbal arguments that had been formulated earlier by Parmenides, Kant, and McTaggart (among others). Gödel sought a strictly rigorous line of reasoning. In the end he found a solution of the Einstein field equations that had the universe rotating in a fashion that mixed up space and time. Gödel's model admitted travel backwards in time. Gödel concluded that, if time travel were possible, then the entire concept of time is impossible. He presented his results to Einstein on the great man's seventieth birthday, along with an etching. Gödel's wife Adele knitted Einstein a sweater for the occasion, but in the end decided not to send it.

Einstein was not pleased by Gödel's contribution to relativity theory. This was a bleak time in the physicist's life. His family situation was unstable and generally unhappy, he had few friends, he had become isolated among physicists because of his rejection of the quantum theory, and his efforts to establish a unified field theory had failed.

In his later years, after Einstein's death, Gödel became ever more isolated. He did, reluctantly, go to Harvard in 1953 to receive an honorary doctorate, and to hear his work on incompleteness praised as "the most important mathematical discovery of the previous 100 years". But he complained of being "thrust quite undeservedly into the most highly bellicose company" of John Foster Dulles (who also received an honorary degree). Gödel refused to go to Washington in 1975 to meet President Gerald Ford and receive the National Medal of Science. He suffered hallucinatory episodes and spoke of certain evil forces in the world "directly submerging the good." He preferred to conduct all conversations by telephone, even if his interlocutor was just a few feet away. He became convinced that there was a plot to poison him, and in the end starved himself to death. The death certificate read that the cause of death was "malnutrition and inanition" brought on by "personality disturbance".

$\mathcal{I}$n many mathematics departments the process for hiring new faculty is an elaborate, Byzantine contrivance. The tradition at UCLA has been to have a 12-person "Staff Search Committee" execute the process (the committee was affectionately known, among the *illuminati*, as "Sniff & Snort"). This committee would spend entire afternoons poring over dossiers, intently reading letters of recommendation, and subjecting them to the most excruciating exegesis. Great discussions, and sometimes arguments, would unfold over what the letter-writer meant by using "that" instead of "which," or by making this binary comparison rather than that one. In the same vein, the University of Washington once received a letter of recommendation for a very fine PhD student who was finishing up at UCLA. The letter waxed fulsome with praise, and added—just as an aside—the fact that the candidate was a talented soccer player. Soon someone from Seattle phoned someone from UCLA and asked, "What is the hidden message in this line about soccer?"

$\mathcal{E}$dgar Lorch (1907–1990) received an offer in 1943 to go to the Institute for Advanced Study and be John von Neumann's assistant. This was quite an honor, but Lorch had misgivings about how much extra work this would entail. He had heard reports of an assistant's duties varying from ε to $1/\varepsilon$, depending on who the master was. So Lorch went to Oswald Veblen (1880–1960), then the Director of IAS, to learn more about his new job. Veblen quickly, and with some annoyance, described four categories of activity:

1. For von Neumann's lectures, take notes, complete proofs, prepare mimeograph sheets of them, distribute them to the auditors.

2. Assist in the editing of the *Annals of Mathematics*, of which von Neumann was the chief editor. Prepare all accepted manuscripts for the printer. [Mark all fonts (Greek, boldface, German, etc.); indicate displayed formulas.]

3. The *Annals* were being printed in the U.S.A. for the first time and no longer by Lütke and Wolf in Nazi Germany. The assistant was to go to Baltimore two afternoons per week to teach the printers how to typeset superscripts, subscripts, etc.

4. At that time von Neumann was still writing up his many 100-page papers in German. The assistant was to translate, type up, and prepare these many papers for publication.

Veblen added firmly that these were the basic and normal duties of the assistant, but other duties could be added as the need arose. John von Neumann was a ball of fire in those days, and his productivity was awesome and prodigious.

As it turned out, Lorch also got the offer of a Cutting Travelling Fellowship which would enable him to go to Szeged, Hungary and study with F. Riesz (1880–1956). He decided that that was the more attractive option, and he declined the offer from the Institute. He later learned that the many "assistant duties" that had been described to him by Veblen had been parceled out to four distinct individuals.

About 35 years ago the *Encyclopedia Britannica* re-invented itself. Rather than provide an alphabetized collection of articles about the human condition, the reference work's new edition would now be organized in broad categories according to subject areas and affinity of ideas. And there was a separate pair of volumes, called the *Syntopticon*, that provided a sort of concordance to the main text. This was to be an entirely new concept in encyclopedia design, and it was introduced with much fanfare. Complex analyst Don Wilken (1938–) of SUNY at Albany fell for it. He bought the leather-bound version of the new work and also invested in the custom walnut cabinet to hold it. Don anticipated that this new acquisition would be a great boon to his childrens' education and that he would enjoy it as well. When it arrived at his house, he set it up in the living room with great anticipation and brought the whole family, and some of the neighbors, in to admire his new prize possession. Next day he sat down and started reading the new *Brittanica*.

After about two weeks, Wilken decided that he hated the damned thing. He did *not* like the organization, he couldn't find the information he wanted, and he wasn't getting anything out of it. His kids found it to be opaque and useless. The more he read, the more irate Don got. He finally decided he'd had enough. He called up all his friends and organized a big outdoor party. At the party, Don built a huge bonfire, gave a flowery speech condemning the encyclopedia to eternal damnation, and ceremonially burned his *Encyclopedia Brittanica*: the books, the leather covers, the custom walnut case—the whole *schmegeggy*. What an impression this made!

Now SUNY at Albany has a grand tradition. Once per year someone holds a big outdoor party and builds a huge bonfire. Everyone brings something in their life that they just hate, and they ceremonially burn it. Often little speeches are made, libations are drunk, and a good time is had by all. It is not known whether anyone has ever sacrificed a car—perhaps a Rabbit or a Yugo??

$\mathcal{W}$e told in *Mathematical Apocrypha* of Alfred Tarski (1902–1983) changing his name from Teitelbaum to Tarski to protect himself from anti-Semitism. People liked to give Tarski a hard time about his name. None other than Paul Erdős tells of Tarski giving a mathematical talk and presenting a particularly nice theorem. Somebody in the audience said, "You have been anticipated by Teitelbaum." Tarski's face grew red, and the vein in his neck pulsed visibily. "*I* am Teitelbaum," he asserted.

$\mathcal{A}$lfred Tarski was consumed by, and was aggressive about, his work. When Leonard Gillman was writing his PhD thesis, he showed some of the results to Tarski. Tarski pointed out that some of the results were related to published work of himself and Erdős. Tarski started leaning on the young man in an alarming way, and it soon looked as though Tarski were proposing a joint paper. Others intervened on Gillman's behalf, and the matter was settled amicably.

But around this time the Gillman family got a dog, and they named it "Tarski." When asked why, Mrs. Gillman explained that "a dog is something that brings both joy and pain." The existence of the Tarski dog became widely known in the logic community, and many jokes were the result.

$\mathcal{H}$ugo Steinhaus (1887–1972) was a prolific mathematician by any measure—also my mathematical great, great grandfather. He had many accomplishments, not least of which was the Uniform Boundedness Principle in functional analysis. But his "greatest discovery" was Stefan Banach (1892–1945). Indeed, one summer evening he was walking in the park in Cracow and he overheard a young man discussing all the latest ideas about modern mathematics. Steinhaus introduced himself and he and the young fellow (Banach) soon became fast friends. Steinhaus got Banach introduced into the fast talk and hot mathematics of the Scottish Cafe in Lwow, and the rest we can read about in the annals of mathematics.

Hugo Steinhaus

Saunders Mac Lane was one of the grand old men of American mathematics. Educated in Göttingen before World War II, a student of Hilbert and Bernays, he has had a huge impact on modern mathematics. He tells this story of the current form of his name:

> I was born nearby in Norwich on August 4, 1909, christened Leslie
> Saunders MacLane. My father and his brothers had changed the
> spelling of the family name from MacLean to MacLane so as not to
> be considered Irish. It was my nurse who suggested the name Leslie,
> but a month later, my parents agreed that they didn't like the name.
> My father put his hand on my head, looked up to God, and said,
> "Leslie, forget." I have gone by two last names ever since. The space
> in Mac Lane was added years later by me, when my first wife,
> Dorothy, found it difficult to type our name without a space.

One of the many major contributions that Saunders Mac Lane made to mathematics was the invention, joint with Sammy Eilenberg, of category theory. Although Eilenberg had pronounced their joint 1945 paper on the subject to be the final word, Mac Lane and many others wrote a good many papers about categories. In 1971 Saunders Mac Lane really put his imprimatur on the subject by writing the book *Categories for the Working Mathematician*. Of course his intention was to make the abstract idea of categories available to mathematicians outside the obvious specialties of homological algebra and algebraic geometry. But he received a certain amount of guff for this title. In fact none other than J. Frank Adams (1930–1989) once spoke of a prospective book entitled *Categories for the Idle Mathematician*.

In the original volume *Mathematical Apocrypha*, we told the tale of Hans Rademacher's (1892–1969) flawed proof of the Riemann hypothesis. In fact it was a disproof, but it was still in error. None other than Carl Ludwig Siegel found the flaw. The event did grace the pages of *Time* magazine, in the April 30, 1943 issue. The article contained a picture of Riemann with the caption "Few understand it, none has proved it." The *Time* reporter wrote

> A sure way for any mathematician to achieve immortal fame would be
> to prove or disprove the Riemann hypothesis ... No layman has ever

been able to understand it and no mathematician has ever proved it.

One day last month electrifying news arrived at the University of Chicago office of Dr. Adrian Albert (1905–1972), editor of the *Transactions of the American Mathematical Society*. A wire from the society's secretary, University of Pennsylvania professor John R. Kline, asked editor Albert to stop the presses; a paper disproving the Riemann Hypothesis was on the way. Its author: Professor Hans Adolf Rademacher, a refugee German mathematician now at Penn.

On the heels of the telegram came a letter from Professor Rademacher himself reporting that his calculations had been checked and confirmed by famed mathematician Carl Ludwig Siegel of Princeton's Institute for Advanced Study. Editor Albert got ready to publish the historic paper in the May issue. US mathematicians, hearing the wildfire rumor, held their breath. Alas for drama, last week the issue went to press without the Rademacher article. At the last moment the professor wired meekly that it was all a mistake; on rechecking, mathematician Siegel had discovered a flaw (undisclosed) in the Rademacher reasoning. US mathematicians felt much like the morning after a phony armistice celebration. Said editor Albert, "The whole thing certainly raised a lot of false hopes."

*P*aul Erdős had a very restless spirit. He rarely stayed more than one month in any one place. He was forever anxious to see what was around the next bend in the river. As his friend and protector Ron Graham put it, "He always knows that there's another problem out there. When he starts making small talk about politics, you know he's wrung the mathematical juices out of the environment. He's ready to move on."

*T*he Web site www5.in.tum.de/~huckle/mathwar.html contains fascinating and detailed information about the imprisonment, torture, and extermination of various mathematicians during the German terror in World War II. As an instance, Stefan Banach's body was used to breed lice in the laboratory of Professor Weigl in an effort to develop an anti-typhus serum. Both Fritz Hartogs and Felix Hausdorff committed suicide in the 1940s (rather than continue life under the Nazi regime). Stanislaw Saks was killed by the Gestapo in 1942. Karol Borsuk, Ernst Hellinger, Alfréd Renyi,

Paul Turán, Egon Balas, Heinrich Grell, Curt Herzstark, and many others were held in concentration camps. Many prominenet mathematicians successfully resisted the War effort. Many others died. It was a dreadful time.

$\mathcal{B}$efore World War II, André Weil used to enjoy touring the ancient sites of Europe. Outside Rome, he was particularly fond of going to concerts at the ancient Roman *Augusteo*. Sadly, this building was destroyed by the fascists. Commenting on the situation, Vito Volterra (1860–1940) said (of Mussolini), "That man has done more damage in Rome than in Madrid."

Vito Volterra

Volterra's prize pupil was Fantappié, up until the day when he came to Volterra to sing the praises of the anti-Semitic legislation that Mussolini (1883–1945) had just introduced in Italy. Of course Volterra was Jewish, and everyone knew it. "How is it possible," he later said, "that I did not have the presence of mind to throw him down the stairs?"

$\mathcal{D}$avid and Gregory Chudnovsky (David was born in 1947, Gregory in 1952) are two mathematicians, Russian by birth, who work out of a small apartment in Manhattan. They have an affiliation (as "Senior Research Scientists") with the Math Department at Columbia University. But this entails no official duties. They spend all their time toiling away in Gregory's apartment. Their efforts are subsidized by the work of their wives, and they support their mother.

Gregory is quite ill—he is crippled up by the rare disease *myasthenia gravis*. He must stay in bed most of the day. But he is a brilliant thinker. His older brother David, who is considerably more ambulatory, is also a gifted mathematician.

One of the Chudnovsky brothers' passions is calculating the digits of π.

They have, with their own hands and with mail-order parts and parts canni-balized from other machines, built a one-gigaflop[*] parallel processing supercomputer. For a time, they had one of the fastest computing machines in the world. They call their machine *m zero* or, more compactly, $m0$. They have calculated more than two billion digits of the irrational number π, and for a time this was the record.

They are, in effect, looking for order in the chaos of the decimal expansion of a transcendental number. So far, they have not succeeded.

The Chudnovsky brothers' laboratory is Gregory's apartment. They keep the place virtually dark, in part because the computer already stretches the electric circuits to the limit and in part because Gregory's eyes cannot tolerate light. They are devoted to each other and to their work. They have an ongoing and furious relationship with Federal Express, as that is the conduit for their never-ending supply of computer parts (and return of non-functioning parts).

Brilliant as they are, and respected around the world, the Chudnovskys have never landed regular academic jobs in this country. They do not fit into the standard paradigm of what is thought to be a productive and functioning mathematician. David is sometimes perceived to be obnoxious and uncooperative. But they are passionate and inspiring scientists, very much in the spirit of Kepler and Newton. It is only too bad that they do not have students so that they can pass on their accumulated knowledge.

$\mathcal{H}$arvey Dubner (1928–) is one of those, along with the Chudnovsky brothers and a few others, who calculate exotic facts about the prime numbers. Here are some of his fairly recent discoveries:

- The integer composed of 2700 digits 1 followed by 3155 digits 0 followed by a single 1 is a prime.
- The known prime with the greatest number of zeros in its decimal representation is $134088 \times 10^{15036} + 1$. This prime integer has 15036 zeros in sequence in its decimal expansion.

$\mathcal{A}$t the end of his life, after John von Neumann became quite ill, he could no longer carry out his duties at the Institute for Advanced Study and cer-

---

[*] Here a "flop" is a floating point operation, the most basic operation of a computing machine.

tainly could no longer do all the traveling and consulting that was once his habit. UCLA put together a special offer for von Neumann, and built a special office for him—complete with built-in plumbing. When I was there in the 1970s, the office still existed and was inhabited by graduate students. The plumbing had been removed.

John von Neumann's final years were very sad. He saw his magnificent mental powers fade almost before his eyes. His friend Eugene Wigner recalls: "When von Neumann realized he was incurably ill, his logic forced him to realize that he would cease to exist, and hence cease to have thoughts .... It was heartbreaking to watch the frustration of his mind, when all hope was gone, in its struggle with the fate which appeared to him unavoidable but unacceptable."

In his final days, von Neumann turned to Catholicism for comfort. The loss of his powers was exceedingly painful for him. Eventually he had a complete psychological breakdown. There was uncontrolled panic, screams of terror each night. His friend Edward Teller (1908–2003) said, "I think that von Neumann suffered more when his mind would no longer function, than I have ever seen any human being suffer."

G. D. Birkhoff was arguably the first great American mathematician. This is because he proved the first results that were esteemed and respected by the great mathematicians of Europe.

And Birkhoff was an ardent defender of American mathematical interests. Stanislaw Ulam (1909–1984) says, a bit petulantly, of Birkhoff: "In discussing the general job situation, he would often make skeptical remarks about foreigners. I think he was afraid that his position as the unquestioned leader of American mathematics would be weakened by the presence of such luminaries as Hermann Weyl, Jacques Hadamard, and others. He was also afraid that the explosion of refugees from Europe would fill the important academic positions, at least on the Eastern seaboard. He was quoted as having said, 'If the American mathematicians don't watch out, they may become hewers of wood and carriers of water.' "

B. L. van der Waerden was a distinguished Dutch mathematician and author of one of the all-time great algebra books [VAN] (that is still in use today). Unfortunately, there is evidence that he was a Nazi sympathizer during World War II. [I should stress that some, such as Horst Tietz, speak very highly of van der Waerden. According to Tietz, van der Waerden was his mentor and ally in resisting the Nazis.] It was made plain to him after the War that he was not welcome to return to The Netherlands.

Stanislaw Ulam was trained as the purest of pure mathematicians. He was part of the cohort that used to hang out at the Scottish Cafe in Lwow and developed many of the early ideas in set theory, topology, and Banach space theory. Later in life, Ulam became very applied: he had some of the key ideas both for the atomic and the hydrogen bombs, and he worked for many years at Los Alamos. He found the transition from pure to applied mathematics to be remarkably easy. The physicist Otto Frisch (1904–1979), on his first visit to Los Alamos from Britain, wrote, "I also met Stan Ulam early on, a brilliant Polish topologist with a charming French wife. At once he told me that he was a pure mathematician who had sunk so low that his latest paper actually contained numbers with decimal points!"

Sophie Germain (1776–1831) was the daughter of a wealthy, middle class silk merchant. She was enchanted by the story of Archimedes (287 B.C.–212 B.C.) being slain by a soldier because he was so absorbed in his mathematics that he would not answer the soldier's queries. For social and personal reasons, Sophie's father opposed her study of mathematics. She was, however, undaunted. She would work at night wrapped in a blanket, because they had taken away her clothing to keep her from getting up. They took away her heat and her light as well. But this only hardened Sophie Germain's resolve. Such was Sophie's dedication to her subject that her father finally gave in and she was allowed to devote herself to mathematics.

After Gauss (1777–1855) published his book *Disquisitiones Arithmeticae*, he enjoyed correspondence with one "Monsieur LeBlanc". Imagine his surprise in discovering that this person was in fact a woman—indeed, it was Sophie Germain. Gauss's letter to Germain, after he discovered her true identity, reveals something about the man:

> But how can I describe my astonishment and admiration on seeing my esteemed correspondent Monsieur LeBlanc metamorphosed into

this celebrated person, yielding a copy so brilliant it is hard to believe? The taste for the abstract sciences in general and, above all, for the mysteries of numbers, is very rare: this is not surprising, since the charms of this sublime science in all their beauty reveal themselves only to those who have the courage to fathom them. But when a woman, because of her sex, our customs and preju- dices, encounters infinitely more

Sophie Germain

obstacles than men, in familiarizing herself with their knotty prob- lems, yet overcomes these fetters and penetrates that which is most hidden, she doubtless has the most noble courage, extraordinary tal- ent, and superior genius. Nothing could prove to me in a more flatter- ing and less equivocal way that the attractions of that science, which have added so much joy to my life, are not chimerical, than the favour with which you have honoured it.

*J*ohn E. Littlewood thought a great deal about what makes for creativity and what are the attributes of a man that most contribute to effective and hard work. Late in life, he wrote a book entitled *The Mathematician's Art of Work*. It contains this rather telling paragraph:

Mathematics is very hard work, and dons tend to be above the aver- age in health and vigor. Below a certain threshold a man cracks up, but above it hard mental work *makes* for health and vigor (also—on much historical evidence throughout the ages—for longevity).

*J*t is not generally well known that the first university in the Western world was in Bologna, founded in 1150. Both the universities of Paris and Oxford were founded about 50 years later, and Cambridge some time after that. Soon after their founding, both Cambridge and Oxford suffered severe problems with discipline among the students. It seems that the gatherings of high-spirited young men, alien to the townfolk and independent in their thinking, led to frequent disorder and sometimes riots. Frequently these riots culminated in a murder. The local magistrates found the situation particularly frustrating because priests and clerks (and, by osmosis, the student body) enjoyed a certain immunity. Bologna suffered the same problem—an offense by one student against another which would ordinarily be punishable by law was in fact immune to the local governance. Oxford and Cambridge endeavored to deal with this situation of lawlessness by putting their students into residence halls, which later became known as colleges. The Master of the College had considerable power over his "kids".

At Cambridge, the second in command, under the Master, was the "Father of the College". He was analogous to what we would now call a Dean. When the degree examinations were held, the Dean would act as a partisan on behalf of the student (against the faculty). If the candidate could not stand up to the close questioning of the examiners, the Dean would step in and speak on his behalf. As a result, nobody ever failed these exams. The really excellent students graduated *summa cum laude*, but everyone got a degree. Because the disputing parties (student and faculty) occupied three-legged stools, the exam came to be known as the *Tripos* (more on this exam in the first volume of *Mathematical Apocrypha*); in memory of the confrontational nature of the exam, the students who passed "first class" came to be known as "Wranglers".

After the Wranglers came the "Senior Optimes", then the "Junior Optimes". When the last Junior Optime—the lowest man on the totem pole—came forward for his degree there was a ceremony by which an enormous wooden spoon was lowered, and he was given this utensil as a consolation prize. The expression "to get the wooden spoon" has thus become proverbial.

*J*n 1799 Thomas Jefferson (1743–1826) received a letter from a young man asking which branches of mathematics would be most useful to study. In reply, Jefferson praised Euclid (325 B.C.–265 B.C.) and Archimedes. He went on to say that trigonometry "… is most valuable to every man. There is scarcely a day in which he will not resort to it for some of the purposes of common life; the science of calculation also is indispensable as far as the extraction of the square and cube roots. Algebra as far as the quadratic equation and the use of logarithms are often of value in ordinary cases; but all beyond these is but a luxury; a delicious luxury indeed; but not to be indulged in by one who is to have a profession to follow for his subsistence …"

$\mathcal{M}$y teacher J. W. T. Youngs (1910–1970) was a rather formal fellow of the old school. He was a really nice guy, but one had to get past his rather austere and gentlemanly *gestalt*. The fact that he had recently, along with Gerhard Ringel, proved the Heawood Conjecture only added to his brash self-confidence.

One day Youngs announced to our class that his fondest desire was to work the word "omphaloskepsis" into one of his papers. Sadly, he learned shortly thereafter that he had a terminal illness. The word appears as the last word in his last paper.[*]

$\mathcal{J}$ames W. Glover (1868–1941) was Chairman of the Department of Mathematics at the University of Michigan in the 1920s. He had an unusual (at least by today's standards) method of recruiting new faculty. He made dittoed copies of a flyer that he sent around to those whom he considered to be promising young mathematicians, inviting them to respond if they felt they might be interested in a position at Michigan. Raymond L. Wilder (1896–1982) tells of his flyer arriving while he was away visiting Ohio State University. Evidently it was thrown on the porch by the mailman and was picked up by his oldest daughter. She was two years of age at the time, not a great reader, and she tore the flyer up into small pieces.

Later on Wilder's wife found the pieces, put them together, realized what the document was, and Wilder *did* write to Professor Glover. He *did* get the job at Michigan, and the rest is history.

$\mathcal{A}$fter Paul Dirac (1902–1984), Nobel Laureate in Physics, retired, he moved to Tallahassee, Florida in order to be near his daughter. De Witt Sumner (1941–) tells me that, as an Assistant Professor at Florida State, he nearly made his mark by mowing down Dirac one day when Sumner was rushing to class in his car and Dirac was painstakingly and slowly trying to cross the road.

---

[*] For the uninitiated: "omphaloskepsis" means "contemplation of one's navel."

Of course Paul Dirac was a great physicist. Being a theoretical type, he liked to theorize about all problems of daily life, rather than to find solutions by experiment. Once, at a party in Copenhagen, he proposed a theory according to which there must be a certain distance at which a woman's face looks its best. He argued as follows: at distance $d = \infty$ one cannot see anything anyway; but at $d = 0$ the oval of the face is deformed because of the small aperture of the human eye, and many other imperfections (such as small wrinkles) become exaggerated. By the intermediate value theorem (or words to that effect), we conclude that there is a certain optimum distance at which the face looks its best.

George Gamow (1904–1968), a Russian physicist, dared ask Dirac, "Tell me, Paul, how close have you seen a woman's face?" "Oh," replied Dirac, holding his palms about two feet apart, "about that close."

Several years later, Dirac married the sister of the noted Hungarian physicist Eugene Wigner. One of Dirac's old friends, who had not heard about the betrothal, dropped in unannounced on Dirac at his home. To his surprise, he found Paul Dirac with an attractive woman who served tea and then sat down comfortably on the sofa. "How do you do?" said the friend, wondering who this comely woman might be. "Oh!"exclaimed Dirac, "I am sorry. I forgot to introduce you. This is … this is Wigner's sister."

J. Robert Oppenheimer (1904–1967) was working in Göttingen when, one day, the physicist Dirac came to him and said, "Oppenheimer, they tell me you are writing poetry. I do not see how a man can work on the frontiers of physics and write poetry at the same time. They are in opposition. In science you want to say something that nobody knew before, in words which everyone can understand. In poetry you are bound to say … something that everybody knows already in words that nobody can understand."

It is said of Paul Dirac that he would only say exactly what he meant to say and nothing more. Once, when someone was making polite conversation at dinner, it was indicated that it was windy outside. Dirac immediately left the table and went to the door. He looked outside, nodded in satisfaction, and returned to the table. "Indeed, it is windy," uttered Paul Dirac. It has been said, presumably in jest, that Dirac's spoken vocabulary consisted merely of "Yes", "No", and "I don't know."

One of the grand old men of mathematics at Columbia University in the 1920s and 1930s was Edward Kasner (1878–1955). A distinguished geometer, he was described fondly by his office mate Edgar R. Lorch. Kasner would arrive each winter day swaddled in topcoat, jacket, and sweater. He would systematically peel them all off, and then replace the jacket. He would straighten his tie and brush himself off in preparation for his impending lecture. The last step of the ceremony was that Kasner would turn his back to Lorch, pull an envelope from his jacket containing his false teeth, and snap them quite audibly into place. Then off to teach.

Kasner gave a popular course for M.A. candidates which was quite elementary. A good part of the course concerned large numbers (integers). One of the great man's favorite numbers was 10^{100}, which is considerably greater than any number which arises naturally in the physical universe. Kasner asked his (some sources say two-year-old and some sources say nine-year-old) nephew what name to give to this monstrous number, and the boy gurgled, "googol." The name has stuck.

$\mathcal{E}$dward Kasner was in the habit of walking up Riverside Drive (from Columbia University) to the New Jersey ferry and then paying the nickel fee to ride across the river. He would then climb the Palisades to the forest on top. This was all part of his regimen for good health and well being. It was his habit, each time he did this, to bury a nickel at the base of some tree. When asked why, he replied that this way he would never find himself without ferry fare on his return.

$\mathcal{J}$ames Victor Uspensky was a Russian by birth. He eventually emigrated to the United States and settled at Stanford University. One day Uspensky, Hans Frederick Blichfeldt (1873–1945), Maxwell Alfred Heaslet (1907–1976), and Bacon were riding in Blichfeldt's car to an AMS meeting in Berkeley. They speculated on the number of colleges, universities, and other organizations that would be represented at the meeting. One guess was ten, another was a dozen. Someone observed that, if the meeting were in New York, then there would be a hundred or so. Uspensky then said very solemnly that "Yes, we must recognize that we live in a remote province."

$\mathcal{H}$alsey Royden (1928–1993) describes Uspensky as "… a kind and gentle, soft spoken man; quite formal in manner. But he liked to seem quite 'tough'—Once I was at his home at a small gathering of graduate students, and he was making a vigorous argument upon some political theme. Suddenly he drew himself up and announced, 'I have Tartar blood in my veins. That is why I am so fierce!' And once he gave me reprints of two of his papers that he had written in Spanish and published in an Argentine journal. I thanked him, but had to confess that I could not read Spanish. 'Well,' said he sternly, 'learn it!' "

$\mathcal{I}$n the first volume of *Mathematical Apocrypha*, we told one story of David Hilbert's struggles with the concept of Hilbert space. Here is another, courtesy of Saunders Mac Lane.

> … J. von Neumann in 1927 introduced the axiomatic description of a Hilbert space, and used it in his work on quantum mechanics. There is a story of the time he came to Göttingen in 1929 to lecture on these ideas. The lecture started "A Hilbert space is a linear vector space over the complex numbers, complete in the convergence defined by an inner product (a product $\langle a, b \rangle$ of two vectors, a, b) and separable." At the end of the lecture, David Hilbert (by custom sitting in the first row of the lecture hall of the Mathematische Gesellschaft), who was then evidently thinking about his definition and not about the axiomatic description, is said to have asked, "Dr. von Neumann, ich möchte gern wissen, was ist dann eigentlich ein Hilbertscher Raum?"[*]

"One can measure the importance of a scientific work by the number of earlier publications rendered superfluous by it."

—David Hilbert

---

[*] Freely translated, this is "Dr. von Neumann, I would very much like to know, what after all is a Hilbert space?"

$\mathcal{M}$arshall Hall, Jr. (1910–1990) tells of political intrigue in the Yale Math Department when he returned from World War II: "I was released from the Navy in 1945 and returned to Yale. Here things had gone badly as there was a feud between Øystein Øre and Einar Hille, who had married Øre's sister. Øre was the Sterling Professor and nothing could be done to him. But his enemies took it out on me, saying that in no circumstances could I be promoted and given tenure. One of the enemies, Nelson Dunford, made the amazing statement that my 'Projective Planes' paper was so good that he doubted that I could write another good paper.

I had two more years to go on my term as Assistant Professor, and I was offered an Associate Professorship at Ohio State. Professor Longley (1880–1965), then Chairman, was able to have me offered the salary of an Associate Professor, but Professors Hille, Dunford, and Wilson would not hear of promotion, to the amazement of the Dean. And so I accepted the Ohio State offer and started a new era."

$\mathcal{M}$ikhael Gromov (1943–) tells of attending lectures about cosmology by two topologists—really great topologists—concerning the possible shape of the Universe. Gromov asked the first of these whether the Universe was simply connected. The man replied, "It is clear that the Universe cannot be but simply connected, for non-simple connectedness would imply some high-scale periodicity, which is ridiculous." The other topologist gave a talk entitled "Is the Universe simply connected?" which seemed to be related to the question. When informed of the first speaker's statement, he said, "Who cares, it's still a meaningful question, like it or not."

In discussing these divergent opinions, Gromov offers his own: "Take a loop in the Universe, a reasonably short loop compared to the size of the Universe, say of no more than 10^{10} to 10^{12} light years long and ask if it is contractible. And, to be realistic, we pick a certain time, for example 10^{30} years, and ask if it is contractible within this time. So you are allowed to move the loop around, say at the speed of light, and try to determine whether or not it can be contracted within this time. The point is, even imagining our space to be some topological 3-sphere S^3, we can organize an innocuous enough metric on S^3 so that it takes more than 10^{30} years to contract certain loops in this sphere and in the course of contraction we need to stretch the loop to something like 10^{30} light years in size. So, if 10^{30} years is all the time you have, you conclude that the loop is not contractible and whether or not $\pi(S^3) = 0$ becomes a matter of opinion."

$\mathcal{J}$n *Mathematical Apocrypha* we spoke of the fact that two of Gauss's sons settled in the St. Louis area. In fact there are now six Gausses listed in the St. Louis telephone directory, and one can see a vanity license plate that says "Gauss" when driving around town.

$\mathcal{A}$ well-known number theorist was asked to be on an editorial board for a new journal. Not having heard anything, the journal contacted him again nine months later. He replied that if they wanted someone as an editor who left his correspondence unanswered for nine months, then he was their man. He became an editor.

$\mathcal{T}$oday it is common for students in good high schools in the United States to learn calculus—while still in high school. The high schools are convinced that this is one essential way in which a high school student can demonstrate ability to do college level work: in order to be competitive for admission to a good college or university, a student must take calculus in high school. Of course there is a price to pay for this acceleration. For example, in recent years, classical Euclidean geometry has been shortchanged in the high schools because of the mad rush to get to calculus.

In fact the more gifted high school students get through all of calculus and even some linear algebra and beyond while still in high school. My own university has two fifteen-year-olds (brothers) who are currently the standouts in our first-year graduate courses!

But things were different years ago. Garrett Birkhoff relates that, "Even in the 1930s, mathematics concentrators (majors) in many small colleges only got to calculus in their senior year!"

$\mathcal{B}$oth German mathematicians F. Hausdorff (1868–1942) and F. Hartogs (1879–1943) committed suicide in the 1930s because they felt that they could not survive in Hitler's Germany. They feared being sent to a concentration camp, and preferred to die with some dignity. In the case of Hausdorff, he had mistakenly thought that his high position in German science would protect him. So he simply waited too long. Hausdorff and his wife and her sister all sat in the living room and took poison. Before they took their lives, the Hausdorffs sent their daughter to a safe abode. She survived the Nazi terror, and lived into her nineties.

*J*n the same vein, Ivan Niven tells of over 100 active mathematicians who came to the United States because of the Hitler terror in the 1930s. Of these, at least 50 were eminent in research. Naturally, such scholars of distinction desired to continue their scientific work at schools that were oriented towards such activities. But creating 50 such professorships at the 25 American schools that had active mathematics research programs at that time was out of the question. The result was that many an outstanding emigré had to take what was available.

Among the noted immigrants to the United States in those days were

From Germany: Emil Artin, Alfred Brauer, Richard Brauer, Herbert Busemann, Richard Courant, Max Dehn, K. O. Friedrichs, G. P. Hochschild, Hilde Geiringer-Pollaczek, Fritz John, Rudolf Carnap, Peter Lax, Hans Lewy, Otto Neugebauer, Emmy Noether, William Prager, Hans Rademacher, C. L. Siegel, Richard von Mises, Aurel Wintner, Hermann Weyl.

From Hungary: Paul Erdős, George Pólya, Tibor Radó, Ottó Szász, Gábor Szegő, Theodor von Kármán, John von Neumann.

From Austria: Kurt Gödel, Karl Menger, Abraham Wald.

From Czechoslovakia: Charles Loewner.

From Yugoslavia: William Feller.

From Poland: Nachman Aronszajn, Stefan Bergman, Salomon Bochner, Samuel Eilenberg, Witold Hurewucz, Mark Kac, Jerzy Neyman, Alfred Tarski, Stanislaw Ulam, Antoni Zygmund.

From Russia: Stefan Warschawski, Alexander Weinstein.

From France: Léon Brillouin, Claude Chevalley, Jacques Hadamard, Raphael Salem, André Weil.

*E*xamples of distinguished foreign mathematicians accepting positions well below their station are:

- Stanislaw Ulam, who had taken his Ph.D. in Poland in 1933, came to the University of Wisconsin as an instructor in 1941.
- Alfred Tarski was appointed to an instructorship at Berkeley in 1942, at the age of 40.
- Richard Courant went to New York University in 1934 as a professor, with an annual salary of $4000, contrasted with his previous salary equivalent to about $12000 as Director of the Mathematical Institute at Göttingen.
- Max Dehn (1878–1952), a distinguished topologist who was the first mathematician to solve one of the 23 famous problems posed by Hilbert at the turn of the century, ... got a position at the University of Idaho, Southern Branch in 1941 at a salary of $1200. Later he moved to a position at the Illinois Institute of Technology.

*T*here are a number of stories about how the famous astronomer Tycho Brahe (1546–1601) met his end. A popular rendition is that Brahe once attended a regal bash. He drank a *lot* of beer. On the return trip from the party, his carriage went over a very large bump. Brahe's bladder burst, and he died.

One common story is that he was attending a dinner party. He needed to urinate, but did not leave the table for many hours out of politeness. He later became quite ill with a bladder infection, and suffered for eleven days before he died. A variant of this view of the matter is that Brahe had a prostate condition that contributed to his demise; this, together with the bladder control situation, led to kidney failure and death.

The most recent theory, offered by Joshua and Anne-Lee Gilder in their new book *Heavenly Intrigue* is that Johannes Kepler (1571–1630, Brahe's famous student) actually *poisoned* Brahe—very slowly—in order to get his data (that he used in order to do the calculations for his celebrated three laws of planetary motion). Evidently Brahe did not want to share his data (collected over a period of more than forty years) because he feared that they would be used to support the Copernican theory. It is a fact that Kepler never got the files from Brahe himself; after considerable haggling, he obtained them from the great man's heirs.

*E*veryone knows that the life of an Assistant Professor is hell. One is working like a devil to establish research credentials, teaching credentials, and an overall positive gestalt in the department. I certainly got little sleep during my probationary period. One day I was giving a lecture in my advanced class on harmonic analysis. I stepped back to have a look at what I had written on the blackboard and I fell asleep—*while standing on my feet!!* Don Marshall (1947—), now a professor at the University of Washington, finally had the temerity to wake me up so I could get on with the lesson.

*A*ccording to Lee Rubel (1928–1995), the University of Illinois Math Department hired 75 people from 1961 to 1962. The hiring in those days was so fast and furious that chairmen would show up at the January meeting of the AMS/MAA with blank offer letters—only a name needed to be filled in. They would stand in the public areas and tap people on the shoulder, asking them if they wanted a job.

But there were at least some standards in those days. Ads used to appear in the *Notices* of the AMS and other venues that said, "Algebraists need not apply."

A student of John Tate (1925—) was giving a seminar at Columbia University, in the days when Serge Lang (1927—) was on the faculty. The poor fellow said something rather foolish, and Lang would not let it go. He kept questioning the speaker, and could not get a satisfactory answer. Lang finally ended up polling each person in the room, asking what they thought of this ridiculous statement.

*F*or a period of several years Lars Hörmander was the Director of the Mittag-Leffler Institute in Djursholm, Sweden. Hörmander is a man who gives passionate attention to all details. He used to mow the lawn at the institute himself (and he would try to shame the visiting members to pitch in). He also would coordinate the many clocks at the institute so that they would all chime at the same time.

*M*y colleague Robert McDowell's wife, Att, has a wry sense of humor. One week Bob was having a guest in—to give a colloquium—and Bob was particularly pleased. This guy was, after all, one of the great authorities in

Bob's field. Bob's eyes were aglow with anticipation, and he told his wife. "Oh," said Att. "Well, Bob, if he were a painter would you say he was like Picasso?" "No," said Bob, "I wouldn't go that far." "Well," said Att. "Would you say he is as great as Kandinsky?" Att smiled. "No," said Bob. "I guess he's not Kandinsky either." "Well," said Att, sharpening the knives, "Would you say he is like Watteau?" "Oh, I don't know," cried Bob. "He is just another great mathematician!"

*I*n the first half of the twentieth century, American mathematics was haunted by a specter of anti-Semitism. It should be understood that anti-Semitism was a part of American society at the time; the sins that were committed in mathematics departments—especially the prestigious Eastern ones—were more often than not institutional rather than individual. Nonetheless, there were problems, and certain individuals were mistreated. Lipman Bers (1914–1993), a Jew from Latvia, tells it this way:

> In many places there was no absolute ban on Jewish professors, but if there was one, the chairman of the department was supposed to worry that there shouldn't be too many, "too many" being usually defined as one more.
>
> Peter Lax ... already published the story of Norman Levinson's (1912–1975) appointment to an assistant professorship at MIT. Levinson, Wiener's favorite student, was a natural for the job, but the "not too many" principle prevented it. The famous British mathematician G. H. Hardy, who was visiting MIT at that time, threatened to disclose on the pages of *Nature* that the initials MIT stand for Massachusetts Institute of *Theology*. Levinson was appointed.
>
> I myself was advised by a well-meaning dean to change my first name. "The second is all right, but the first ..." and he suggested the name Lesley. He also advised me to join the Unitarians.

*I*n *Mathematical Apocrypha*, we asserted that Fields Medalist Alexandre Grothendieck (1928–) had disappeared; nobody knows where he is. Meanwhile, we have heard from Roy Lisker (1938–) who asserts that he knows people who can locate Grothendieck in the Pyrenees. Indeed the Web site www.grothendieck-circle.org has all the latest information. It seems that *Notices of the AMS* Associate Editor Allyn Jackson knows how to contact Grothendieck. Her articles [JAC] provide some of the latest information on the great man.

Alexandre Grothendieck

Mathematical Apocrphya} also related the story of Grothendieck turning down the $270,000 Crafoord Prize in 1988 (coincidentally the year of his retirement from his position at IHES at the age of 60). In doing so, he said that he did not need the money and that he had left the world of mathematics in 1970. He went on to say that, "the only decisive proof of the fertility of ideas or of a new vision is that of time. Fertility is recognizable by offspring, not by honors."

In fact Grothendieck expressed considerable anger in his letter to the Swedish Academy declining the prize:

> [T]he ethics of the scientific profession (especially among mathematicians) have degraded to such a degree that pure and simple theft between colleagues (especially at the expense of those who have no position of power to defend themselves) has almost become the general rule and is in any case tolerated by all, even in the most flagrant and ubiquitous cases.

The letter was eventually published in the Parisian newspaper *Le Monde*. Grothendieck went on to say that to participate in the game of accepting prizes and honors would be to validate "a spirit and an evolution in the scientific world that I see as profoundly unhealthy, and condemned to disappear soon, so suicidal is it, spiritually as well as intellectually and materially."

*I*n 1983–1986 Grothendieck penned the 2000-page tome *Récoltes et Semailles: Réflexions et témoignage sur un passé de mathématicien.** This work contains tales of his life, and considerations of his new ideas about mathematics and philosophy. Roy Lisker has translated the first part of the book into English, and details may be found at the URL

---

* Loosely translated, this is *Harvests and Sowings: Reflections and Bearing Witness on the Life of a Mathematician.*

`www.fermentmagazine.org/rands/recoltes1.html`

Grothendieck's own description of *Récoltes et Semailles* is as follows:

There are many things in *Récoltes et Semailles*, and different people will no doubt see in it many different things: a *voyage* to the discovery of a past; a *meditation* on existence; a *portrait of the morals* of a milieu and of an era (or the portrait of an insidious and relentless sliding of one era into another …); an *inquest* (almost detective-style at times, and at others bordering on cloak-and-dagger fiction set in the underbelly of the mathematical megapolis); a vast *mathematical ramble* (which will leave more than one reader in the dust …); a practical treatise on applied psychology (or, if you like, a book of "*psychoanalytic-fiction*"); a panegyric on *self-knowledge*; "My *confessions*"; a private *diary*; a psychology of *discovery and creation*; an indictment (pitiless, as is fitting), even a *setting of scores* in "the world of elite mathematics" (and without any gifts).

In 1987–1988 Grothendieck gave us another book, *La Clef des Songes ou Dialogue avec le Bon Dieu*. He offers therein his reasons for believing that God exists, and that He speaks to people through their dreams.

In fact, since the time he quit research mathematics around 1970, Grothendieck has penned many long manuscripts. Covering topics ranging from Galois theory to "anabelian algebraic geometry" to homotopical algebra to topos theory, these works constitute a remarkable legacy. Much of it has only been circulated privately. Rumor now is that Grothendieck has taken a vow of silence; we do not know whether that vow extends to his writing.

*G*rothendieck's mathematics has had a powerful and incisive impact in part because of his use of abstraction. In fact his point of view has so permeated modern thought that it is difficult to imagine pre-Grothendieck mathematics. But it must be noted that he never pursued abstraction for its own sake. For Grothendieck, abstraction was a means to see things more clearly and to penetrate more deeply.

In a mathematical conversation about number theory, Grothendieck's interlocutor said, "Let us look at a particular prime." Grothendieck was abashed. "What?" he said. "You mean consider a *specific* prime number?" "Yes," said his friend. "OK," replied the great man. "Let's take 57." Someone later commented that surely Grothendieck knew that $57 = 3 \cdot 19$ is not prime. "No," said David Mumford. "Grothendieck never thought in concrete terms."

*J*n the mid-1960s the *St. Louis Post-Dispatch* had a scurrilous reporter—whose politics were slightly to the right of reactionary—who was constantly raking up the muck. He used his column as a tool to attack anyone whom he suspected of leftward leanings.

On one occasion, Herbert Aptheker (1916–2003) and his daughter Bettina (1944–)—well-known rabble-rousers from the U. C. Berkeley Free Speech Movement days—came into town late at night on a train. It happens that this was the only mode of transportation available to them, and they had come to St. Louis to visit relatives. But this journalist wrote up the occasion to the effect that they had snuck into town in the middle of the night, they had come as "outside agitators", and were going to corrupt the minds of our youth.

Around the same time, Washington University in St. Louis had a tenure candidate in one of the departments who happened to be a communist. The case became a *cause célèbre*, and was hotly debated in the faculty club and at departmental teas and also in the *Post-Dispatch*. Guido Weiss of the Washington University Math Department was an ardent defender of this candidate, and so were several other of the mathematicians. Our hardy and right-leaning journalist weighed in on the matter, accusing the university of being led by godless heathens, representatives of the anti-Christ. No verbiage was too tawdry for this member of the fourth estate to use in his arguments.

When, finally, the candidate *did* receive tenure, a big party was held in his honor. The journalist showed up, shook everyone's hand, slapped a lot of backs, and wished everyone well. Quite an about face. By the way, the name of this journalist was Pat Buchanan (1938–).

A ndré Weil, the celebrated algebraic geometer, is said to have given Claude Lévi-Strauss (1908–) a key idea for how to use algebraic structures in anthropology. In fact Weil wrote a chapter entitled "Sur l'étude algébrique de certains types de lois de mariage (système Murugin)" in Lévi-Strauss's book *The Elementary Structure of Kinship*.[*] The issue at hand was to link processes and states normally understood as cultural to those formal pat-

---

[*] In English, the title of this chapter is "On the algebraic study of certain types of laws of marriage (the system of Murugin)."

terns of interaction between individuals or institutions known as social networks. Weil describes the algebraic structure of certain types of marriage laws.

$\mathcal{G}$erhard Kohlmayr (1930–) is a mathematical amateur of some talent who has taken it upon himself to prove that Zermelo-Fraenkel set theory (ZF) is inconsistent. His papers are formulated quite aggressively, and usually point out that the mathematical community has been rather closed-minded and unwilling to give his ideas due consideration.

The Kohlmayr paper that I refereed followed his standard format: he formulated a perfectly sensible, technical theorem about mathematics (in this case operator theory). Then he proved it was true, and he proved it was false. The only possible conclusion, of course, is that ZF is inconsistent.

Let me stress that Kohlmayr's papers—at least the one that I refereed—contain good, solid mathematics. Everything makes sense and is written in the accepted mathematical argot. BUT: the paper that I refereed contained a mistake. Kohlmayr forgot to check that a certain operator was closed. So one of his proofs was invalid.

Kohlmayr is no wilting flower when it comes to trumpeting the significance of his work. In a letter to the editor of *The Pacific Journal of Mathematics* accompanying his paper "Zermelo-Fraenkel Axiomatic Set Theory is Inconsistent," he says, "By this time, there are about a dozen proofs for the inconsistency of *ZF*. But practically everybody thinks I am a fool or worse. Of course, there are a few people, very few indeed, who had doubts about the reals. Among the living are Heyting, Troelstra, Lorenzen and Laugwitz. Possibly Alonzo Church." The closing paragraph of this paper reads

> The preceding two proofs for the inconsistency of Zermelo-Fraenkel set theory are not the only ones. I have worked out several others. We might thus be tempted to conjecture that there is no consistent axiomatic set theory which admits infinite sets. Of course, this is in marked opposition to Cantor's bold insistence on the actual infinite and contrary to the accepted mathematical principles of the day. But then, we should also bear in mind that it might take a bold step to remove the deadwood from mathematics.

I submitted my report to editor Richard Arens (1919–2000). Arens communicated my thoughts to Kohlmayr and rejected the paper. But Kohlmayr would not take this lying down. He wrote back a rather aggressive letter to Arens saying, "Yes, it's true. There may be a technical mathematical error. But it doesn't matter, because Zermelo-Frankel set theory is inconsistent."

*K*arl Marx (1818–1883) is famous for being the father of communism. He spent years during his youth living in the library and formulating his thoughts that were ultimately enshrined in *Das Kapital*. Less well known is that Marx also wrote a calculus book. After all, he had to feed his family.

Utter Credulity

*W*e had a job candidate once who was quite a hit around the department. He charmed everyone's socks off, and we were about ready to hand him the offer on a silver platter. The time came for his ceremonial lecture, and we all clambered into the lecture room in eager anticipation.

The talk began at 4:35 PM, and was scheduled to run until 5:30 PM. It went swimmingly, peppered with interesting mathematics, new ideas, and good stories. The time rolled around to 5:25 PM, and the speaker frowned. He looked at the audience and said, "I need some additional time to finish telling you my ideas." So he walked across the stage and pushed the big hand of the clock back to 5:00 PM. Then he *actually used the additional 25 minutes!* There was no further discussion of this man's candidacy.

*P*aul Sally is a terrific mathematician who suffers from diabetes. He has unfortunately lost an eye, and had portions of his leg amputated (in a sequence of surgeries) because of his illness. But Paul forges on. He recently won the Franklin and Deborah Tepper Haimo Teaching Award from the MAA for the fantastic outreach program he has set up with the Chicago schools.

One of Paul's students, Sarah Drucker, learned the Banach-Tarski paradox from him. Roughly speaking, the paradox says that you can take a solid ball of radius one, disassemble it into finitely many pieces, and re-assemble them into two balls of the same original radius (of course the pieces are non-measurable and the construction requires the dreaded Axiom of Choice). More generally, you can take almost any solid, disassemble it into finitely many pieces, and re-assemble it into any other solid. The student was so inspired that she wrote an homage to Paul inspired by the Banach-Tarski paradox—the message was that Paul had been disassembled and re-assembled and he was now more than the sum of his parts. Sarah's poem was ultimately published as *My Proof of the Paradox*:

My Proof of the Paradox
for Paul Sally, Professor of Mathematics

I have torn apart a thousand oranges, searching for the unmeasurable
set, looking for the five easy sections I could rotate and reglue
to get a citrus ball the size of the sun

looking for the Banach-Tarski subsets of the physical world,
the purely theoretical in something I could touch. You taught me
the Axiom of Choice, used it to find me a set without measure,
played a few tricks with rotation matrices, and POW, the Paradox.

It would be better than Physics, better than Chemistry, better
than Alchemy, if it worked. Three years I sat in your classes,
three years I doubted it. Three years I watched diabetes

pull you apart: first your eye, then your foot, then a leg.
But Choice, like any Axiom, relies on faith, and you
must have found some physical cadenza inside your body,
you must have found a way to reglue, so

you were not only my teacher, you were my Proof
of the Paradox. You were not dying, but growing, choosing
to grow into a blaze
bright enough to illuminate
even the most unlikely mathematics,

but still mysterious,
still just out of reach,
just beyond choice.

*P*aul Erdős's favorite question for anyone was "What's your profession?" He once approached the celebrated Hungarian poet, Ferenc Juhász (1928–), with this query. At that time, Juhász was at the peak of his career, and he was tremendously famous. His writings and his photograph appeared everywhere. His signature moustache was instantly recognizable by one and all. But Erdős never read newspapers or magazines, nor did he watch TV. He was trenchantly unaware of who was famous or who was not. Somewhat abashed at Erdős's query, Juhász replied (somewhat proudly),

"I am a poet." Paul Erdős then innocently asked, "And how do you make a living?"

*J*n 1966 Richard Feynman was slated to appear in a forthcoming Swedish encyclopedia. The editors wanted to show the human side of the great scientist, and somehow soften the effect of all the technical physics that was being explained. They had heard that Feynman played the bongo drums. So they wrote to him and requested a photograph of Feynman playing the bongos.

He wrote this reply:

Dear Sir,

The fact that I beat a drum has nothing to do with the fact that I do theoretical physics. Theoretical physics is a human endeavor, one of the higher developments of human beings, and the perpetual desire to prove that people who do it are human by showing that they do other things that a few other humans do (like playing bongo drums) is insulting to me. I am human enough to tell you to go to hell.

*R*ichard Courant was a fine mathematician but remarkable in many other ways. He founded the mathematics institute in Göttingen. When he had to leave Germany, because of the Hitler terror, he moved to New York and founded the mathematics entity that came to be known as the Courant Institute. André Weil claims that he used to entertain the thought that God, in His wisdom, one day came to repent for not having had Courant born in America, and He sent Hitler into the world expressly to rectify this error. After the war, when Weil shared this thought with Ernst Hellinger (1883–1950), the latter said, "Weil, you have the meanest tongue I know."

*C*omplex analyst Donald Eisenman was an Assistant Professor at the University of North Carolina at Chapel Hill. He was a talented guy, and most likely would have achieved tenure in that excellent department. But he got distracted by politics. He joined the Communist Worker's Party and became involved in fighting the Ku Klux Klan. When there was an armed confrontation with the Klan in Greensboro, North Carolina in the 1970s, Eisenman was there (on the side of the good guys). He was wounded in the shoulder.

*J*n 1984, E. M. Stein led a group of harmonic analysts who spent a month in China giving lectures and interacting with the Chinese graduate students and mathematicians. Of course the Chinese were gracious hosts and treated us to much fine touring and several banquets. At one of the latter events, a lovely Chinese lady came up to me and asked what was the correct response to give in America when someone paid you a compliment. I tend to be a minimalist at these things, so I advised her that the correct thing to say was, "Thank you." She adopted a pensive look, smiled, and said, "That's very interesting. The gentleman across the room told me to say, 'Why, sir, you flatter me too much.'"

One might have been tempted to say that the South will rise again. But one held one's peace.

*J*ean Dieudonné, the great raconteur of twentieth century French mathematics, tells of a custom at the École Normale Supérieure in France to subject first-year students in mathematics to a rather bizarre rite of initiation.

A senior student at the university would be disguised as an important visitor from abroad; he would give an elaborate and rather pompous lecture in which several "well-known" theorems were cited and proved. Each of the theorems would bear the name of a famous or sometimes not-so-famous French

Jean Dieudonné

general, and each was wrong in some very subtle and clever way. The object of this farce was for the first-year students to endeavor to spot the error in each theorem, or perhaps not to spot the error but to provide some comic relief. In the 1930s, some of the students at the École Normale banded together and decided to put together a modest little work that would systematically describe and develop all the key ideas in modern mathematics. They decided to ascribe their efforts to the French general Nicholas Bourbaki, and thus the idea for the famous "Bourbaki" books was born.

*J*t is inevitable that a mathematician will ponder the nature of creativity, how one gets hold of problems and cracks them, what is the nature of the mathematical art. J. E. Littlewood once said that you need to know a problem as well as you know your own tongue going round your mouth. Michael Atiyah (1929–) said,

> It is hard to communicate understanding because that is something you get by living with a problem for a long time. You study it, perhaps for years, you get the feel of it and it is in your bones. You can't convey that to anybody else. Having studied the problem for five years you may be able to present it in such a way that it would take somebody else less time to get to that point than it took you. But if they haven't struggled with the problem and seen all the pitfalls, then they haven't really understood it.

*T*here was a period in the 1960s and 1970s when Soviet scientists were under tremendous pressure to compete with scientists in the West. This was a time of fast and furious publication, and many unsubstantiated claims. Alexander Ivić (1949–) describes one incident of the politicization of mathematics:

> There was a mathematician in the Soviet Union … called Nikolai Gavrilov, who in the early 1960s was a big Communist boss in one of the Ukrainian cities. He was an amateur mathematician, and he got hold of the Riemann Hypothesis and thought he had a proof, and being a man of power he organized for his proof to be published. The fortunate thing is that the proof did get published. You should know that at that time the Soviets were competing, especially with the United States, in all fields—the cosmos, high technology, and so on—and they had very fine number theorists and mathematicians in general. Especially one Gelfond and one Yuri Linnik. Now, when these men got hold of a false proof of the Riemann Hypothesis, it is said, Gelfond had a heart attack and Linnik was choking and gasping for breath. He almost died from shock. They were so frustrated

because much damage had been done to Soviet mathematics by this false proof of the Riemann Hypothesis appearing.

*J*n the late 1970s at UCLA there was a very special logic seminar called the "Cabal Seminar". One might wonder about the provenance of this unusual name. Certainly it suggests something dark and mysterious for the cognoscenti. It turns out that the seminar was named after their favorite real estate agent. Whenever they used her services to help a new mathematician relocate, she would give them a kickback from her commission. And they used the money to run the seminar.

The pleasures derived from this largesse were quite evident. On Fridays, when the rest of us were at tea eating Ritz crackers and drinking tepid tea, the logicians would be sitting off in the corner drinking wine and eating camembert and pâté de foie gras. And they were able to bring in a number of classy speakers for their mathematical activities. The permanent record of this seminar was published by Springer-Verlag, with editors consisting of various subsets of A. S. Kechris (1946–), D. A. Martin (1940–), and Y. N. Moschovakis (1938–).

A friend of mine, who used to teach at the University of Calabria in southern Italy, had the following astonishing experience. As will soon become clear, names must be omitted in order to protect the guilty. For convenience, we shall call my pal Marco.

This chum went to the library—in the days before *MathSciNet*—in order to look up a certain paper. He pulled the appropriate volume of *Math Reviews* off the shelf and began to page through it. When he came to the correct juncture, he found that the desired page was neatly missing. Hmmm.

The following week my friend happened to be in Padua. He went to the library to find the article in *Math Reviews* that he needed. Once again the relevant page was missing.

A while later he was in Rome. Went to the library. Got the book off the shelf. The page was missing.

After several similar mishaps, he decided that the hard-copy *Math Reviews* had been shipped in a bundle from Providence, Rhode Island— and Italy had received a bad lot. Made sense—no? So he wrote to an

American friend and asked for a photocopy of the relevant page. Soon enough it arrived. And then all was clear.

For the upper left-hand corner of that special page contained a review of a paper by one of Marco's colleagues in Calabria which read, "This author has kindly translated my paper, word for word, from the original Romanian and rendered it in Italian. He has even been so kind as to reproduce several misprints from the original."

After a bit of investigation, it became clear to Marco that the miscreant had ridden a train, equipped with a razor blade, throughout Italy—neatly excising the dramatic page from each copy of *Math Reviews*. But now he had been caught. A major scandal ensued. The villain was roundly and publicly condemned by the president of the Italian Mathematical Union and was ejected from the Society. But he retained his tenured position— some things, after all, are sacred.

*S*aunders MacLane recalls his days (1972—1974) as President of the American Mathematical Society. One day there was lively debate at a meeting of a new bylaw which specified that a certain action could be taken by the Society only with a 2/3 majority. Naturally some wit had to ask, "Why two-thirds." Secretary Everett Pitcher (1912–) was the institional memory for the group and often its strongest voice of reason. He said, "Because it is the simplest number between one half and one." Discussion thereupon ceased.

*I*n his new autobiography, Saunders MacLane tells lovingly of his days with the National Academy of Sciences, and his involvement with many important committees of that august organization. In particular, he served for a time on the National Science Board. One of their memorable vignettes concerned some materials called MACOS (Man, A Course of Study) that were prepared in the 1970s by social scientists to augment the high school curriculum. The purpose of MACOS was to show students how people in other cultures really lived. There was, for instance, material on Eskimo life, and it related how the Eskimos had the tradition of leaving aged and superannuated members of their group on an ice floe to die.

Some parents were vigorously upset by this description of how one might treat granny, and objected strenuously to their congressmen.

Saunders was involved in getting the situation straightened out, and having the offending passages modified.

*A*s we know from *Mathematical Apocrypha*, the concept of "Erdős number" was invented in 1969 by Caspar Goffman GOF]. This numerical invariant measures your "distance" from Paul Erdős by way of intermediary collaborators.

Just so, there is now a concept of "Bacon number." What could this be? Kevin Bacon (1958–) is a hard-working and prolific movie actor who seems to have made a movie with practically everyone. So your Bacon number is 1 if you have made a movie with Kevin Bacon. Your Bacon number is 2 if you have made a movie with someone who has Bacon number 1. And so forth.

It is quite difficult to find anyone with Bacon number greater than 3. The Web site www.cs.virginia.edu/oracle/ will assist you in your search. You will find that Jack Nicholson has Bacon number 1. Laurence Olivier has Bacon number 2. So do James Cagney and Drew Barrymore. Douglas Fairbanks has Bacon number 3. So does Mary Pickford (arguably the very first movie star). Interestingly, Paul Erdős has Bacon number 4. Is it possible? The provenance is this:

Paul Erdős was in *N is a Number* (1993) with Gene Patterson;

Gene Patterson was in *Box of Moon Light* (1996) with John Turturro;

John Turturro was in *Cradle Will Rock* (1999) with Tim Robbins;

Tim Robbins was in *Mystic River* (2003) with Kevin Bacon.

The "average Bacon number" is about 2.941. There are 7452 actors with Bacon number 5, 944 with Bacon number 6, 107 with Bacon number 7, and 13 with Bacon number 8. But it turns out that the more senior actor Sean Connery has a better record. The "average Sean Connery number" is about 2.706. It turns out, however, that actor Rod Steiger is the best "center" of the Hollywood universe. His average is about 2.6519.

*T*he powerful AMS tool MathSciNet now allows you to calculate the <*name*>-number (just like an Erdős number) for any mathematician <*name*>. You simply click the <Authors> button, enter a name, click the <MR CD> button, enter a second name, and the chain of papers connecting the two mathematicians is produced. An alternative route is to go to the URL

`www.ams.org/msnmain/cgd/index.html?sourceid=71496`
There is one pair of mathematicians with degree of separation 14. With the great flowering of mathematical collaboration today, you will find that most mathematicians are separated by a chain of 5 or fewer. Certainly our friend Paul Erdős played an important role in keeping these numbers low.

J am one of the Managing Editors of *The Journal of Mathematical Analysis and Applications*. When papers come into the journal, I assign them to an Associate Editor, whose job it is to obtain a referee's report. Once that report is rendered, it is passed along to the editorial office (located in an office building in San Diego, California). The report is evaluated and a decision made about publication. The editorial office then notifies the author of the fate of his work. All very standard and very simple.

So recently I received an e-mail from an author in Bulgaria who claimed that he had received an e-mail from me in October of 2002 telling him that his paper had been accepted, but that he still awaited the formal letter of acceptance. This he needed because his tenure case was coming up.

Well, I dutifully (but somewhat doubtfully) checked. We had no record of this paper. My archival copy of old e-mails revealed no such correspondence. I thus reported that I was very sorry, but we had no record of his paper nor of our erstwhile putative communications.

Next day I received an e-mail from his supervisor asserting that they really needed a letter from me about the acceptance of this paper—such documentation was required in order to process the scholar's promotion in Bulgaria. After several exchanges, the supervisor was good enough to send along an electronic copy of my e-mail to the author from October of 2002. I was anxious to read it.

To my very great surprise, it was written with a marked and heavy-handed Bulgarian accent (my ethnic background is Hungarian and Sicilian and French, but I am a native speaker of English). This clearly was not penned by me. Thus the jig was up. I reported the crime back to the supervisor, and the fate of our Bulgarian friend is in the offing. Last I heard, the candidate had filed suit against the university and the supervisor had gone on long-term leave.

J recently got a phone call from a fellow who told me that he had 6 (count 'em!) PhD degrees and a lot of great ideas about number theory. Would I listen? Sure, why not. Then he said, "It usually takes me five or six hours to explain my ideas to a math PhD" I said, "Fine. You've got ten minutes."

So he started in. I'm no expert on number theory, but I can prove Dirichlet's theorem and the prime number theorem. I figured I could follow at least some of it. And indeed I could. He talked about some elementary congruences and relations among numbers. There were a few leaps of faith, but on the whole I got the general drift of what he was about. Then he said, "It's as though the number 4 is floating in a pool of oil, and it is attracted by magnetic waves to the number 2." I cleared my throat. He went on to observe that if you imagined all these integers planted on a Möbius strip then you could discern a lot of new information about genetic splicing.

Suffice it to say that we parted friends, but none the wiser.

Penn State University (formerly known as "The Pennsylvania State University," up until the time when they won the national football championship) is a fine university located out in the middle of nowhere. Of course it is in Pennsylvania, and that is a good start. But, for political reasons, it could not be located too close to Philadelphia nor too close to Pittsburgh. So it was instead located in a small town that is right at the intersection of the two diagonal segments of this more-or-less rectangular state.

The town is called "State College," which is already the cause of considerable consternation and confusion. There are debates every few years about changing the name; but so far it has stuck. Penn State began in the mid-nineteenth century as an *agricultural high school*. Over time it evolved to an agricultural college and, eventually, in the early 1960s, to a research university. Today it is a fine school with many strong programs.

But even in the early 1960s the town had many unusual features. At colloquium dinners, the Math group could go to any restaurant in town, have a fine meal, and just tell the waiter to "charge it to the Math Department." No money ever changed hands.

You could walk into any of the retail establishments in State College and find a stack of blank checks on the counter. After selecting your items, you grabbed a blank check, filled in the name of your bank, filled in your name, and wrote the amount of the purchase. Worked like a charm.

Even more charming was the fact that every store had two sets of prices—one for the townfolk and one for the professors.

Yet another anomaly in the town of State College is that there is a tax on *your profession*. The way it works is that each profession is assigned a certain number of *points*, and a point is worth so many dollars. And that is how much tax you pay each year. An Associate Professor rates more points than an Assistant Professor, and a full Professor rates more points than an Associate Professor. It turns out that the profession with the greatest number of points is the football coach—Joe Paterno. That with the least number of points is housewife.

J was a job candidate at Penn State University in 1980. The Chairman, Don Rung, made a great show of treating me like a prince. After my colloquium, he and a group of departmental luminaries took me for dinner to *The Tavern*, a popular local eatery.

When we ordered, Don said "In honor of Professor Krantz, let's order a nice bottle of wine." He peered at the wine list and told the waiter he wanted a particular bottle of Chianti. The young fellow replied, "Actually, I'd recommend this Burgundy over here. "No," said Don. "We want the Chianti." The waiter departed, and we returned to our conversation. Presently the waiter returned and said, "We're out of the Chianti. Can I recommend the Burgundy?" "No," said Don. "Let's have this Merlot." The waiter replied, "I'd really recommend this Burgundy." "We want the Merlot," Don insisted. Off went the waiter.

After a while the waiter returned and informed us that they were out of the Merlot. He suggested the Burgundy. "No, no," said Don Rung. "In honor of Professor Krantz we will have champagne. Go get the champagne." "Well," ruminated the waiter. "I think you would get more enjoyment out of this Burgundy." "Champagne," said Don Rung. The waiter left with a grim look on his face. It did not quite match the grim look on Don Rung's face. After a while the young server returned and said, "Sir, the only wine we have is this Burgundy."

We drank Burgundy that night.

*M*ost mathematicians consider themselves lucky if they produce perhaps 50 papers and a book during their lifetimes. Those among us who pen more than a hundred papers are few indeed. Those who write more than a couple of books are rather rare.

In the mid-1990s, the Institute for Scientific Information (ISI) conducted a study to determine who were the most productive and prolific scientists in the world. As you can imagine, fields like chemistry and physics—where there are often whole teams of people toiling away in a laboratory and grinding out papers at a furious pace—dominated the results. The list of those scientists who produced the most papers during the period 1981–1990 is as follows:

Name, Field, Nation	No. of papers, 1981–1990	Days Between Papers
1. Yury T. Struchkov, chemistry, USSR	948	3.9
2. Stephen R. Bloom, gastroenterology, UK	773	4.7
3. Mikhail G. Voronkov, chemistry, USSR	711	5.1
4. Aleksandr M. Prokhorov, physics, USSR	589	6.2
5. Ferdinand Bohlmann, chemistry, Germany	572	6.4
6. Thomas E. Starzl, surgery, USA	503	7.3
7. Frank A. Cotton, chemistry, USA	451	8.1
8. Julia M. Pola, histochemistry, UK	436	8.4
9. Robert C. Gallo, cell biology, USA	428	8.5
10. Genrikh A. Tolstikov, chemistry, USSR	427	8.6

Suffice it to say that mathematicians are quite a bit further down the food chain.

*W*olfgang Pauli (1900–1958) was more gifted than most physicists at penetrating the mathematical subtleties connected with the study of sub-atomic particles. But one day he was studying a paper by a colleague, and he became more and more befuddled by the complexities of the reasoning. After a while he paused and said, "This isn't right!" Then he thought a while longer and declared, "This isn't even wrong!"

*O*ne of the more remarkable mathematical families of modern times is the Browder family. Earl Browder, the father, was several times the American Communist Party candidate for President of the United States. He did not win.

Earl had three sons, all of them distinguished mathematicians. Felix Browder (1928–), formerly of the University of Chicago and now of

Rutgers, has won the National Medal of Science. William Browder is Professor of Mathematics at Princeton and Andrew Browder is Professor of Mathematics at Brown.

$\mathcal{A}$ lbert Tucker (1906–1995) was one of the grand old men of the Princeton Mathematics Department. He was there in the 1930s, when the Institute for Advanced Study first opened (and shared space with the University Math Department in Old Fine Hall), and he knew all the big names—von Neumann, J. Alexander (1888–1971), Church, Lefschetz (his thesis advisor), and Einstein. He tells many tales of life in the old days. "Einstein didn't wear socks. And James Alexander walked around in tennis shoes—shocking attire by '30s academic standards." It seems that the special shoes enabled Alexander, who was a world-renowned mountain climber, to scale the exterior wall of Palmer Lab when that Physics Department building was locked. Alexander would leave his second-floor office window unlocked in order to facilitate his comings and goings.

As noted in *Mathematical Apocrypha*, Oswald Veblen was one of the leaders of the Princeton Math Department in those days. He was very tall,

Princeton University Department of Mathematics, 1951. Front row: A. W. Tucker, E. Artin, S. Lefschetz, A. Church, W. Feller. Back row: J. T. Tate, J. W. Tukey, D. C. Spencer, R. C. Lyndon, V. Bargmann. Absent: S. Bochner, R. H. Fox, N. E. Steenrod, E. Wigner, S. S. Wilks

so had special jackets made that had an extra (fourth) button. Everyone made jokes about this fact, and the students wrote a verse about it (see *Mathematical Apocrypha*). Veblen was so taken by the English customs of his wife that he was moved to introduce the afternoon tea ceremony. Of course this is one of the hallmarks of mathematical life at Princeton (both at the University and the Institute for Advanced Study, as well as at the Institute for Defense Analyses—in Princeton one can spend all afternoon going to various teas).

Tucker had special memories of Lefschetz, who was his teacher. He says that Lefschetz called Veblen "Uncle" just because Veblen had a habit of name-dropping his famous Uncle Thorstein's name. He says that Lefschetz would heckle even von Neumann during his lectures. And he counseled Einstein, who was constantly harassed by autograph-seekers, to get a haircut as a disguise.

Tucker recalls that, when he was a graduate student at Princeton, Palmer Lab closed at 6:00pm on weekdays and was closed altogether on weekends. But that's where the math library was! If a graduate student wanted to access the books, he/she had to apply for a key. Tucker recalls going to a certain office for this task. He recounts that a "pleasant fellow" requested a dollar deposit, put the dollar in a cigar box, checked off Tucker's name, then got a key for him from another cigar box. All very quaint. Turns out that that pleasant fellow was Karl Taylor Compton (1887–1954), then the Chairman of the Physics Department and soon to become the President of MIT.

The academic year 1969–1970 was a bellwether year in the University of California at Santa Cruz Mathematics Department. During that 12-month period, about two thirds of the math faculty got divorced. Even some people on sabbatical got divorced. I was a tyro at the time—certainly a sycophantic follower of my senior mentors on the faculty. I concluded that what mathematicians did was prove theorems and get divorced. I decided I could live with that.

Erdős related that, from a very early age, he had an ingrained spirit to resist pressure to be like others. He recalls one specific event from his early years:

> I remember an incident when I was a small child. You know, the Jews
> in Hungary had lots of problems after the communist revolution in

1919. There were a lot of anti-Semitic acts. Being a Jew, my mother once said to me, "You know the Jews have such a difficult time, shouldn't we get baptized?" I told my mother, "Well, you can do what you please, but I remain what I was born." It was very remarkable for a small child—I was only six or seven then—because, actually, being Jewish meant nothing to me. It never did.

Tom Jech (1944–) is a gifted and prolific logician at Penn State. He has written some beautiful books about set theory and the Axiom of Choice. In the early 1980s, Tom became a fanatic runner. He ran ten miles at a time—*twice per day*. [Bear in mind that this is the regimen that 4-time Boston Marathon Winner and twice Olympic Gold Medalist Frank Shorter followed when he was training.] In fact he found that he spent the rest of the day sitting in the coffee room talking about running; he was too tired to do anything else.

Well, almost anything else. In point of fact, during the same period, he developed his famous system for predicting the outcomes of football games. The technology was based on integrating over previous won/loss outcomes—studying the resulting graph. The final theorem, on which Jech's predictor was based, hinged on a fairly deep fixed-point theorem.

Of course Penn State is a famous football school. So there was considerable interest in what Tom Jech had to say about predicting the outcomes of football games. He was interviewed on a number of radio shows. Interest in the method waned when it was found that its predictive value was slight. The final nail in the coffin was hammered when Jech admitted in one of his radio interviews that he had never seen a football game.

The *Guinness Book of World's Records* contains a citation for the largest number (i.e., positive integer) ever created by a man. That man is the ineffable Ron Graham, and his number (now known as the "Graham number") is so large that it cannot be expressed with ordinary mathematical or scientific notation. Indeed, Donald Knuth has invented a new scientific argot for describing the Graham number.

Let us first say what the number is, at least in intuitive terms. We seek a positive integer N with the following property. For a set S with N elements, consider all possible subsets (the power set). Now form a new set T

consisting of all pairs of elements of the power set. Partition T into two subsets A and B. We seek the least N so that there are four subsets of S with the properties that **(i)** every pair of these four lies in A or every pair lies in B, **(ii)** every element x that lies in one of these four subsets actually lies in an even number of the subsets.

It was originally suspected that $N = 6$ will do. Now it is known, thanks to a recent result of Exoo [EXO], that N must be at least 11. Graham showed that there is some such number N, and that its value will not exceed the enormous number that we are about to describe.

We define a new type of exponentiation. Let $3 \uparrow k$ denote 3 to the power k. This is a familiar idea. Now let $3 \uparrow \uparrow 3$ denote $3 \uparrow (3 \uparrow 3)$. In general

$$\underbrace{3 \uparrow \uparrow \cdots \uparrow \uparrow 3}_{k \text{ arrows}}$$

denotes

$$\underbrace{3 \uparrow \uparrow \cdots \uparrow \uparrow}_{(k-1) \text{ arrows}} (\underbrace{3 \uparrow \uparrow \cdots \uparrow \uparrow 3}_{(k-1) \text{ arrows}}).$$

Now let

$$M = \underbrace{3 \uparrow \uparrow \cdots \uparrow \uparrow 3}_{3 \uparrow\uparrow\uparrow\uparrow 3 \text{ arrows}}.$$

Next define

$$P = \underbrace{3 \uparrow \uparrow \cdots \uparrow \uparrow 3}_{M \text{ arrows}}.$$

Now iterate this construction 61 more times. You will then obtain Graham's number G.

W. D. MacMillan (1871–1948) came to Chicago as a mature adult—in order to sell his cattle. After the sale, he went to the famous Yerkes Observatory to see the stars through the telescope. MacMillan was so inspired and fascinated that he stayed on to get an education at the University of Chicago. He went through the entire system, earning *summa cum laude* at every stage. Then he remained at the University of Chicago as a faculty member. One day, in his course on potential theory, MacMillan wrote an important partial differential equation on the board with obvious pleasure. He wrote the partial differentiation signs with a great flourish. Standing back to admire the equations, he declared, "That is just beautiful. People who ask, 'What's it good for?', they make me tired! Like when you show a man the Grand Canyon for the first time and you stand there as you do, saying nothing for a while." One could easily see that MacMillan was envisioning the Grand Canyon. "Then he turns to you and asks, 'What's it good for?' What would you do? Why, you would kick him off the cliff!" And MacMillan then kicked a chair halfway across the room.

$\mathcal{A}$s we know, the subject of calculus was beset by confusion and doubts in its early days. One of the main issues was rigor, in particular it was never clear what a limit (obversely, an infinitesimal) was. Bishop Berkeley (1685–1753) was one of the outspoken critics. One of his more pithy broadsides was

> All these points, I say, are supposed and believed by certain rigorous exactors of evidence in religion, men who pretend to believe no further than they can see. That men who have been conversant only about clear points should with difficulty admit obscure ones might not seem altogether unaccountable. But he who can digest [infinitesimals] need not, methinks, be squeamish about any point of divinity.

$\mathcal{T}$he Institute for Defense Analyses (IDA) in Princeton, New Jersey is staffed by many mathematicians—some of them my graduate school classmates. Of course most of their work is classified, so an ex-hippie like myself has no chance of knowing about any of it. But the common wisdom for many years was that the Vietnam War strategy of playing Wagner from low-flying helicopters to terrorize the Viet Cong was hatched at IDA. If you view the Francis Ford Coppola film *Apocalypse Now* then you can see a dramatic representation of this rather extraordinary wartime exploit.

$\mathcal{J}$ean Dieudonné describes the philosophy for what is proper grist for the Bourbaki books as follows:

> … those which Bourbaki proposes to set forth are generally mathematical theories almost completely worn out already, at least in their foundations. This is only a question of foundations, not details. These theories have arrived at the point where they can be outlined in an entirely rational way. It is certain that group theory (and still more analytical number theory) is just a succession of contrivances,

each one more extraordinary than the last, and thus extremely anti-Bourbaki. I repeat, this absolutely does not mean that it is to be looked down upon. On the contrary, a mathematician's work is shown in what he is capable of inventing, even new stratagems. You know the old story—the first time it is a stratagem, the third time a method. Well, I believe that greater merit comes to the man who invents the stratagem for the first time than to the man who realizes after three or four times that he can make a method from it.

Joe Kohn of Princeton was Bergman's Assistant at Stanford in the 1950s. He tells that Bergman was quite taken with Sputnik and all the excitement that went with it. He listened raptly to all the news coverage. One day he came to Kohn and said, "I think that the American people are ready to hear about complex two-space. Phone up NBC (the television station) and tell them that I need three hours per week." Kohn had a tough time weaseling out of this task. But he still loves to tell the tale.

Stefan Bergman

The last twenty-five or so years have seen great growth and productivity in the field of "applied mathematics." This was spurred in no small part by government funding policies. For years mathematicians had struggled with even defining what applied mathematics is. They figured, "Well, if we can't define it then we surely don't have to worry about it."

Paul Halmos went so far as to write an article entitled "Applied mathematics is bad mathematics." He is quick, in the text of the article, to apologize, saying that he only used this title to grab the reader's attention.

Nowadays most Math Departments—at least in the United States—have vigorous applied math groups. Many very pure mathematicians—such as myself, with no training in applied math and little knowledge of applications—find themselves collaborating with scientists from the applied world.

Yet we all can ponder somewhat wistfully the words of G. H. Hardy, in his *Mathematician's Apology*, when he considered the question of applied math.

> I have never done anything "useful." No discovery of mine has made, or is likely to make, directly or indirectly, for good or ill, the least difference to the amenity of the world.

$\mathcal{A}$t the Bologna International Congress in 1928, the meetings began in Bologna and ended in Florence. That is about a three-hour train ride; a special train was engaged to ferry the mathematicians back and forth. Hadamard was placed in a compartment with a particularly noisy bunch of mathematicians; he himself was tired and wanted to take a nap. He was inclined to simply pull rank and ask them to settle down—but they were a bunch of excitable Italians and probably would not have listened. So he instead posed a mathematics problem to the group. Soon the compartment was magically quiet, as everyone was working on the problem. Thus Hadamard took his nap.

"Now, really, these French are going too far. They have already given us a dozen independent proofs that Nicolas Bourbaki is a flesh and blood human being. He writes papers, sends telegrams, has birthdays, suffers from colds, sends greetings.

And now they want us to take part in their canard. They want him to become a member of the American Mathematical Society (AMS). My answer is 'No'." This was the reaction of J. R. Kline, secretary of the AMS, to an application from the legendary Nicolas Bourbaki.

$\mathcal{L}$eonard Roth (1904–1968) was a student at Cambridge in the early twentieth century. He says that one day his director of studies said to him, "I see that Mr. [Leopold Alexander] Pars (1896–1985) is giving a course on general dynamics. I think you might attend." Roth notes that, in those days, everyone was a "Mister." No Cambridge man would have been caught dead with a PhD. Roth had no choice: he had been commanded to take Pars's course, and take it he did. Pars was one of the Cambridge "lightning performers." His course proceeded at blinding speed, and Roth had a terrible time getting anything out of it.

Roth tells that the Tripos consisted of twelve papers, each of duration 3 hours—for a total of 36 hours. He recounts, later in life, that he could not recall any of the 36 hours in detail, except to say that the process was a continuous nightmare. His examinations were held in the great hall of King's College, under the shadow of the famous Gothic chapel. He tells of one rather eccentric individual who wore his gown—as was required—to the exam; but he didn't bother to change his slippers. The officials on duty declined to admit him because he was improperly dressed. It is unknown whether he was able to rectify his haberdashery and return to the exam.

$\mathcal{I}$ was once a guest at the Math Department of Brown University. I stayed in a hotel just down the road from the math building—evidently the canonical place for visitors. When I was checking out of the hotel, the hostelier said to me, "I see you are a mathematician. It looks like you are a pure mathematician. I've always wondered what the difference is between a pure mathematician and an applied mathematician." [Of course Brown has a separate Applied Mathematics Department, so the man running the hotel would be acutely aware of the distinction between pure and applied.] I said to him, "Applied mathematicians get paid more." He smiled and thanked me for this useful information.

$\mathcal{I}$n the late 1970s, David Gillman (1938–) of UCLA got a grant from the Regents of the University to support an innovative project in mathematics instruction. He wrote three short screenplays and hired actors from Hollywood. The end product was three short, and rather remarkable, vignettes to illustrate ideas from probability theory. One of these is a musical about venereal disease—this to illustrate Bayes's theorem. A second is about an African American client of an auto insurance company protesting his rates—and specifically claiming that he is being penalized for being black. Another is a confrontation between an Army Colonel and a mathematician to determine the likelihood that the United States will be annihilated by a nuclear war. The films are instructive and entertaining and (to some people) a little perturbing.

*B*enjamin Peirce, after whom the Peirce Assistant Professorships at Harvard are now named, was a powerful influence on American mathematics. He helped shift the emphasis from teaching and service to research. He published a number of important texts, and was considered to be an impressive and inspiring teacher. He was described by Charles Eliot, later President of Harvard, in this way:

> His method was that of the lecture or monologue, his students never being invited to become active themselves in the lecture room. He would stand on a platform raised two steps above the floor of the room, and chalk in hand cover the slates which filled the whole side of the room with figures, as he slowly passed along the platform; but his scanty talk was hardly addressed to the students who sat below trying to take notes ... No question ever went out to the class, the majority of whom apprehended imperfectly what Professor Peirce was saying.

Another student, later to become a famous mathematician, wrote, "Although we could rarely follow him, we certainly sat up and took notice. I can see him now at the blackboard, chalk in one hand and rubber in the other, writing rapidly and erasing recklessly, pausing every few minutes to face the class and comment earnestly, perhaps on the results of an elaborate calculation, perhaps on the greatness of the Creator."

*R*aoul Bott, one of the most accomplished mathematicians of our time, laments his envy of J.-P. Serre (1926–): "Serre is a prime example of what I call a 'smart mathematician'—as opposed to a 'dumb one.' What he knows is so crystal clear in his mind that he can give us lesser mortals the feeling that it is indeed all child's play. He also had, and still has, the infuriating habit of never seeming to work....[He is] ... never in the sort of mathematical fog so many of us inhabit most of the time. If one asks him a question, he either knows the answer immediately, and then inside out, or he declines comment. I would ask him: 'Well, have you really thought about this?' He would say, 'How can I think about it when I don't know the answer?' Added in Proof: Serre disputes this interchange and asserts that it is just one of the many untruths in my 'pack of lies.'"

*H*alsey Royden tells of makeshift arrangements in the Stanford Math Department in the early 1950s:

> Gábor Szegő arranged for me to be a teaching fellow at Stanford for the summer term of 1950, and I earned my keep by teaching the three calculus classes that were offered that term. Visiting mathematicians for the summer included M. Fekete (1886–1957), W. Rogosinski (1894–1964), W. Fenchel (1905–), and Walter Hayman. Hayman and Fenchel were accompanied by their families. Szegő solved the housing problem for them by renting a fraternity house for the summer and putting them all there. One morning when Rogo met Fekete as they were both shaving, he asked Fekete if he had spent a good night. Fekete replied, "Not bad. I proved the following theorem…"

*O*ne of the more remarkable figures in the history of American mathematics was Nathaniel Bowditch (1773–1838), a native of Salem. Bowditch had to leave school at the age of 10 to help his father in the cooperage business. He was therefore almost entirely self-educated. After teaching himself Latin and reading Newton's *Principia*, he sailed on ships to take four round trips to the East Indies. He then published *the* most widely used book on navigation, *The New American Practical Navigator*. After that achievement, he became an executive actuary for a series of insurance companies.

Harvard University awarded Bowditch an honorary M.A. in 1802 and then offered him the Hollis Professorship of Mathematics and Philosophy in 1806—this to a man who had never gone to high school or college! Bowditch declined the prestigious offer because he could not raise his growing family on the salary offered ($1200 per year). He remained an actuarial executive. Bowditch was, however, a prominent member of Boston's American Academy of Arts and Sciences, and he stayed quite active in Harvard affairs.

Bowditch became best known in American scientific circles for his translation of Laplace's *Mécanique Céleste*; he provided copious notes explaining many sketchy derivations in Laplace's original. Bowditch did this work around the same time that Robert Adrain (1775–1843) showed that Laplace's value of 1/338 for the earth's eccentricity should actually be 1/316. Bowditch decided (correctly) that it is nearer to 1/300.

*O*ne of the traumas of postwar America was the McCarthy era. During this bleak period of the 1950s, there was a national phobia of communism and communist sympathizers. A great many irrational and unfair accusations and threats were made, and many innocent people were persecuted. W. S. Massey (1920–) tells this story:

> There was one other national trauma which had an effect on American mathematicians of my generation: the McCarthyism of the late 1940s and early 1950s. For a few mathematicians, the effects of McCarthyism were absolutely devastating. One of the most brilliant of my fellow graduate students at Princeton, who got his PhD degree in the late 1940s, was unable to get a job for several years during the 1950s. Nobody ever accused him of being a Communist, or of even being subversive, but no institution dared to hire him because of the politics of his father. Later, after McCarthyism died out, he became a full professor at one of our most prestigious universities. However, most of us survived the McCarthy era with minimal damage. But the events of that period remind us vividly that anti-intellectualism abounds, and that the academic world is always dependent on the good will of the proverbial "man in the street" for its survival.

*M*athematicians Lee Lorch (1915–) and Chandler Davis, both mathematical men of strong principle, were persecuted during the fifties because of repressive policies during the McCarthy era. Davis actually served time in prison. They are both noble spokesmen for justice in and beyond mathematics; their tribulations seem to have made them stronger.

*C*handler Davis, who suffered personally under the McCarthy repression, tells it this way.

Between 1947 and 1960 it was even harder than usual for left-wingers in the United States to get by. If you were active on the left, or were thought to be, there were more ways then than now that you could be arrested or threatened with arrest, or have civil rights such as the right to travel abroad withdrawn. But for the would-be mathematicians, the punishment for leftism was meted out not only by government but also by university administrations. You could lose your job, or be passed over for a job; even at the student stage, you could lose a fellowship or in rarer cases be expelled from school.

For me and many of my contemporaries, these were lessons we imbibed, not exactly with our mothers' milk, but with, say, the Plancherel Theorem.

$\mathcal{C}$handler Davis goes on to tell the legendary story of Dirk Struik: "Dirk J. Struik (1894–2000) was 'investigated' by state agencies, leading to his indictment in 1951 for (among other such counts) conspiracy to overthrow the Commonwealth of Massachusetts. MIT suspended him with pay for the duration of his court case, and when it was finally thrown out in 1955 (because the Supreme Court ruled that another state's similar anti-subversion statute was unconstitutional) Struik returned to regular duties. Not all major universities let pass such opportunities to punish people for charges the courts had thrown out.

Dirk Struik

$\mathcal{D}$avid Widder of Harvard University describes his experience (when he was a student) with various famous professors:

"I had Modern Geometry under Bôcher (1867–1918). In the first weeks he had us discovering properties of the ellipse from familiar ones for the circle by use of affine transformations. This was just a foretaste of the marvels to come. I think it was the influence of this course by this instructor that determined for me the choice of a career. In any case I determined to take any course Bôcher offered in later years. The same year I studied Calculus under another famous mathematician, teaching from his own text. Professor W. F. Osgood had a less inspiring style. I recall that he gave us good advice, ignored by most, on how to prepare a paper. You were to fold it down the middle, put a first draft on the right, corrections on the left. He used rubber finger caps to hold chalk. On the whole I would describe him as somewhat imperious."

$\mathcal{A}$ndré Weil reminisces warmly about his student days. He says that Jacques Hadamard is the man who shaped him into a mathematician. But he attended classes with other professors, including Henri Lebesgue (1875–1941) at the Collège de France. During the winter, the classroom was occupied both by fervent students and by the neighborhood drunks— the classrooms were mandated to be open to anyone. So the vagrants took the opportunity to warm up. Weil relates that the drunks had no compunctions about dozing off, but they were not indifferent to Lebesgue's lecture. They never stayed long enough to hear about measure theory. The students would make bets on how long the guests would last. None ever held out for longer than eight minutes.

$\mathcal{D}$avid Widder has his own story along these lines: "If Osgood was imperious, Goursat (1858–1936) was regal. An usher opened the door for his entrance and escorted him out at the close. He lectured in a vast amphitheatre, nearly filled (some said partly by the street people who came in for warmth), and had absolutely no contact with his audience. Although I had taken many French courses in college I still had trouble following the lectures. It was only near the end of the year that I became at all at ease with the language."

$\mathcal{O}$n some additional Harvard professors Widder had this to say: "Of course I had taken several of Birkhoff's courses. His style of teaching was very different from Bôcher's. He presented a view of a research man at work. He would sometimes give the appearance of solving a problem for the first time, with no fear of being stuck, as stuck he sometimes was. But he would tackle the same problem at next lecture and eventually solve it. We learned by trying to understand. O. D. Kellogg (1898–1932) was a more polished, if less memorable, lecturer. J. L. Walsh (1895–) was more like a colleague, for he ate with a few of us teaching fellows in Memorial Hall. He could be very formal in class, very informal after hours. I recall one hot spring day when he gave his lecture on partial differential equations on the front steps of Jefferson Hall.

$\mathcal{T}$homas S. Fiske (1865–1944) describes the role of Charles S. Peirce (1839–1914) in the New York Mathematical Society (later to become the American Mathematical Society).

> Conspicuous among those who in the early nineties (i.e., 1890s) attended the monthly meetings in Professor Van Amringe's lecture room was the famous logician, Charles S. Peirce. His dramatic manner, his reckless disregard of accuracy in what he termed "unimportant details," his clever newspaper articles describing the meetings of our young Society interested and amused us all. He was advisor of the New York Public Library for the purchase of scientific books and writer of the mathematical definitions in the Century Dictionary. He was always hard up, living partly on what he could borrow from friends, and partly from odd jobs such as writing book reviews for the *Nation* and the *Evening Post* .... At one meeting of the Society, in an eloquent outburst on the nature of mathematics, C. S. Peirce proclaimed that the intellectual powers essential to the mathematician were "concentration, imagination, and generalization." Then, after a dramatic pause, he cried: "Did I hear someone say demonstration? Why, my friends," he continued, "demonstration is merely the pavement upon which the chariot of mathematics rolls."

$\mathcal{E}$mil Artin was a dashing and forceful figure in the Princeton Math Department. He was held in awe by faculty and students alike. Gian-Carlo Rota, an undergraduate at Princeton in the 1950s, describes him in this way:

> A great many mathematicians in Princeton, too awed or too weak to form opinions of their own, came to rely on Emil Artin's pronouncements like hermeneuts on the mutterings of the Sybil at Delphi. He would sit at tea time in one of the old leather chairs ("his" chair) in Fine Hall's common[s] room, and deliver his opinions with the

abrupt definitiveness of Wittgenstein's or Karl Kraus's aphorisms. A gaping crowd of admirers and worshippers, often literally sitting at his feet, would record them for posterity. Sample quips: "If we knew *what* to prove in non-Abelian class field theory, we could prove it"; "Witt was a Nazi, the one example of a clever Nazi" (one of many exaggerations). Even the teaching of undergraduate linear algebra carried the imprint of Emil Artin's very visible hand: we were to stay away from any mention of bases and determinants (a strange injunction, considering how much he liked to compute). The alliance of Emil Artin, Claude Chevalley, and André Weil was out to expunge all traces of determinants and resultants from algebra. Two of them are now probably turning in their graves.

*S*heldon Axler (1949–) of San Francisco State University perhaps decided to carry on Emil Artin's paradigm when he penned a linear algebra book a few years ago. His original working title was *Down With Determinants.* The publisher and the editors decided that this title was too unseemly, so they finally devolved upon the title *Linear Algebra Done Right.* Some punctilious folk tried to talk Sheldon out of this title (suggesting "Linear Algebra Done Correctly"), but to no avail.

*H*assler Whitney (1907–1989) reminisced that he once visited Hodge in Cambridge. "In our taking a walk together, he said 'Lefschetz claimed to have proved that theorem before I did; but I really did prove it first; besides which the theorem was false!'...."

Whitney goes on with thoughts of Lefschetz. "With both Alexander and Lefschetz in Princeton, they naturally had many discussions on topology. But Alexander became increasingly wary of this; for Lefschetz would come out with results, not realizing they had come from Alexander. Alexander was a strict and careful worker, while Lefschetz's mind was always full of ideas swimming together, generating new ideas, of origin unknown."

*E*dward Thorp (1932–) was a mathematician at the University of California at Irvine—indeed, he was one of the founding members of the Math Department (the campus opened in 1965). About forty-two years ago he wrote a book [THO] that made him famous, and also made him a fair bit of money. It was called *Beat the Dealer* [THO], and it described a counting system for winning at the Blackjack tables in Las Vegas. His system really worked, and they had to change their *modus operandi* in Vegas to accommodate those who used Thorp's system.

Thorp subsequently wrote a second book, co-authored with Emeritus Professor Sheen T. Kassouf (1928–), called *Beat the Market* [KTH]. This work explained how to use similar methods to beat the stock market. One measure of the success of this technique was that Thorp subsequently opened a stock trading office in Newport Beach, California. He went on to found Edward O. Thorp Associates, and for many years ran one of the most successful hedge funds in the country.

Thorp was Professor of Mathematics at Irvine from 1965 to 1977. He was Professor of Mathematics and Finance from 1977 to 1982. He has a proprietary investment strategy that has been quite successful, and he is now a wealthy man. He and his wife recently founded the Edward and Vivian Thorp Endowment in the Department of Mathematics at U. C. Irvine. Karl Rubin holds the Chair Professorship that they endowed. The endowment money is being managed according to Thorp's investment strategy.

*O*f course businesses make their way in the world by buying and selling. They view everything as a commodity. Thus it should come as no surprise that AT&T once tried to patent the Cauchy integral formula. As a complex analyst, I can only breathe a sigh of relief that they failed.

*M*y friend John D'Angelo (1951–) was a student of Joe Kohn at Princeton. One day he went to Joe with one of his half-baked ideas. He went on at some length about how the idea ought to work. What was lacking was a *bona fide* proof. After a while Kohn fixed him with a stern look and said, "You remind me of the guy who said, 'I've invented a perpetual motion machine. I just need something that goes like this.'" as he pinched his fingers in and out in a periodic motion.

*J*oe Kohn of Princeton University was a student of D. C. Spencer. They were very close. [Kohn was, in turn, one of my teachers.] One day Kohn and Spencer were out at a restaurant having lunch, and Spencer proposed that they celebrate by having a martini. Kohn demurred. "If I drink a martini then it puts me to sleep." "You're lucky," declared Spencer. "With me it takes six."

*J*oe Kohn likes to say that there is no such thing as strong coffee. There are only weak men.

*T*he remarkable book *Codebreakers* [COD], about the cryptological exploits of Arne Beurling (1905–1986) during World War II, tells many fascinating tales of the rather enigmatic life of this great mathematician. Beurling and Lars Ahlfors were great friends, and their exploits carousing and drinking were legend. The book tells particularly of many a late evening in which Ahlfors, Beurling, and Mrs. Erna Ahlfors would all end up in bed together. Everything was quite chaste. One need only imagine the three of them, often fully clothed, laid out side by side.

*M*athematician Leonid Vaserstein (1944–) emigrated to this country in the dark old days of the Soviet Union, when emigration by Jews was quite tricky. Usually the candidate snuck in through Israel, and it was all done in a cloud of secrecy.

Vaserstein's first parking place was the University of Maryland. It was prudent for Vaserstein to keep his arrival in the United States quiet, so most people were unaware that he was in residence in College Park. One day there was a seminar being given by a mathematician who had heard nothing of Vaserstein's emigration; but the topic of the talk was one on which Vaserstein was an expert. And Leonid was in the audience.

As the speaker began, he soon was quoting a theorem of Vaserstein. This small, bearded man in the back raised his hand and said, no, that was not what Vaserstein had proved. The theorem actually said such and such. The speaker acknowledged his gratitude and went on. After a while he cited another Vaserstein result, and the dark, mysterious figure in the back again corrected him. This situation repeated itself several more times. Finally the speaker, in both awe and exasperation, threw up his hands and exclaimed, "How could you possibly know all these things?" In a deep and mysteriously accented voice the man in the back said, "*I* am Vaserstein."

*C*harles Proteus Steinmetz (1865–1923) was mathematician, inventor, and electrical engineer. He pioneered the use of complex numbers in the study of electrical circuits. After he left Germany (on account of his socialist political activities) and emigrated to America, he was employed by the General Electric Company. He soon solved some of GE's biggest problems—to design a method to mass-produce electric motors, and to find a way to transmit electricity more than than three miles. With these contributions alone Steinmetz had a massive impact on mankind.

Steinmetz was a dwarf, crippled by a congenital deformity. He lived in pain, but was well-liked for his humanity and his sense of humor. He was certainly admired for his scientific prowess. The following Steinmetz story comes from the Letters section of *Life* Magazine (May 14, 1965):

> Sirs: In your article on Steinmetz (April 23) you mentioned a consultation with Henry Ford. My father, Burt Scott, who was an employee of Henry Ford for many years, related to me the story behind the meeting. Technical troubles developed with a huge new generator at Ford's River Rouge plant. His electrical engineers were unable to locate the difficulty so Ford solicited the aid of Steinmetz. When "the little giant" arrived at the plant, he rejected all assistance, asking only for a notebook, pencil, and cot. For two straight days and nights he listened to the generator and made countless computations. Then he asked for a ladder, a measuring tape and a piece of chalk. He laboriously ascended the ladder, made careful measurements, and put a chalk mark on the side of the generator. He descended and told his skeptical audience to remove a plate from the side of the generator [at the marked spot] and take out 16 windings from the field coil at that location. The corrections were made and the generator then functioned perfectly. Subsequently Ford received a bill for $10,000 signed by Steinmetz for G.E. Ford returned the bill acknowledging the good job done by Steinmetz but respectfully requesting an itemized statement. Steinmetz replied as follows:

> - Making chalk mark on generator $1
> - Knowing where to make mark $9,999
> - Total due $10,000

For many years the mathematics building at the Swedish Royal Institute of Technology (the Kungliga Tekniska högskolan, affectionately known as KTH) was world famous. It seems that, inside, it resembled the cell block from the Bette Davis/Spencer Tracy movie *Twenty Thousand Years in Sing Sing*—multiple stories, with jail cell piled upon jail cell upon jail cell. Thus this eminent Swedish Math Department became known as "Sing Sing." It was quite common for a mathematician to say, "I'm going to spend my sabbatical at Sing Sing"—and everyone would know what he was talking about. Sadly, the Math Department at KTH has now moved to a new building.

There is one particular professor at KTH whom the secretaries and staff all refer to as "Jesus." It seems that they believe he exists, but they have never seen him.

Russ Hall (1949–) was the math editor for the Penn State University Press. Russ's lifelong dream had been to be a writer—a serious writer. After graduating from college, he was even in a writer's workshop at the University of Iowa.

He lived for several years in a coldwater flat in New York City peddling his wares; he lived the life, ate cold beans and crackers, and tried to break into the business. At one stage he wrote poetry under the *nom de plume* of "R. Yazum Hall." His book of poems won him the prize of "Most Promising Young Black Poet in America". Unfortunately Russ is not black. When he was asked to send his photo in to the awarding agency, the jig was up (and the agency not very pleased). Russ had to relinquish the prize.

I am happy to say that Russ is now a full-time writer of fiction, and he is prospering. He lives outside of Austin and spends his time fishing and writing.

A member of the Southern Illinois University Medical School went one day to a furniture store to buy a new sofa. She selected one for $1200, and was pleased to note that it was on sale for a 15% discount if the buyer was willing to pay cash. So she said to the clerk, "My purchase comes to $1020." The clerk looked up in astonishment and exclaimed, "Oh my God. You're an *idiot savant*."

Paul Nevai of Ohio State University is a Hungarian mathematician who follows in the tradition of Gabor Szegő. Like this author, he is an ardent partisan of Hungary; but he is rather more knowledgeable. In celebration of his heritage, and of mathematical culture in general, Paul composed the following quiz:

Paul Nevai's Quiz on Hungarian and Mathematical Culture

QUESTION #1. True or false: one of Max Born's grandchildren is much more famous than he is. I mean, incomparably more famous, leaves him in the dust and even more.

QUESTION #2. Who is the most famous mathematician buried in the city which, among others, used to be called Leningrad?

QUESTION #3. True or false: the person buried nearby Euler is way more famous than he is.

QUESTION #4. True or false: if the answer to question #3 is "true," then is the person uniquely defined?

QUESTION #5. Who is the most famous mathematician born in the city which, among others, used to be called Leningrad?

QUESTION #6. On the grand scale of things, which mathematician is more likely to be remembered in the 4th millennium, Euler or the one from question #5?

QUESTION #7. Who is the most famous mathematician who was born less than 1 km away from the place where I was born? [*Hint:* The word "place" means the house of the parents, not the hospital.]

QUESTION #8. Who is the most famous physicist who was born less than 1 mile away from the place where I was born? Is he defined uniquely? [*Hint:* The word "place" means the house of the parents not the hospital.]

QUESTION #9. Who is the most famous person who was born less than 1 km away from the place where I was born? [*Hint:* "place" means the house of the parents not the hospital.]

QUESTION #10. Is Peter Lax more famous than his uncle? Is he also richer?

QUESTION #11. If we pick the best soccer player mathematician and the best soccer player physicist in the history of mankind, then the brother of which one is more famous?

QUESTION #12. Who is the most famous mathematician buried in Columbus, Ohio?

QUESTION #13. Who is the most famous mathematician buried in Bloomington, Indiana?

QUESTION #14. Who is the most famous mathematician born in Columbia, Missouri?

QUESTION #15. Who is the most famous mathematician born in the Polish equivalent of Göttingen?

QUESTION #16. Who is the most famous mathematician whose co-author killed more people than the Unabomber?

$\mathcal{N}$ear the end of his career, R H Bing flew to a large American city to attend a conference. As luck would have it, he found a colleague with whom to share a taxi to the hotel. They chatted about this and that, and Bing waxed philosophical about being a grand old man in his field. He said, "My big worry is that the Poincaré conjecture will be proved while I am still alive, but I won't be able to understand the proof."

Answers to Nevai's Quiz

1. Olivia Newton-John. **2.** Leonhard Euler. **3.** See the answer to #4. **4.** Not unique. Choose from among Dostoevsky, Tchaikovsky, Mussorgsky, Rimsky-Korsakov. **5.** Georg Cantor. **6.** TBA on 1/1/3001. **7.** John von Neumann. **8.** Eugene Wigner or Edward Teller or Leo Szilard. **9.** Either Theodore Herzl or Harry Houdini. **10.** The uncle is Gabor Szegö. **11.** Harald Bohr has a more famous soccer-playing brother than Niels Bohr. However, Niels's brother was an even better soccer player than Harald's brother. **12.** Hans Zassenhaus. **13.** Max Zorn. **14.** Norbert Wiener. **15.** Stanislaw Ulam. **16.** Either Hadamard or Mittag-Leffler or Pólya or Henri Cartan. The killer was André Bloch.

Utter Confusion

$\mathcal{T}$here is a custom at Harvard to post the (written) qualifying exams after they are given. That way the members of the department can enjoy the problems and discuss them. One year, Jean-Pierre Serre was visiting Harvard at this time, and he loitered in the hall one day reading the algebraic topology exam (this is a topic about which Serre, a Fields Medalist, knows quite a lot). One of the exam questions had two parts:

1. Define the higher homotopy groups.
2. Calculate the higher homotopy groups of the sphere S^3.

Serre said, "First, I define the higher homotopy groups to all be zero." He paused. Then he said, "Well, what do you know? All the higher homotopy groups of S^3 are zero!"

$\mathcal{A}$t the University of Texas in Austin, Bruce Palka and his wife went with John Tate and his wife to a performance of the play *Proof.* Leaving the show, they entered an elevator in the parking garage. Another couple from the play was there, and the woman said to her husband, "Just what is a Germain prime?" Tate blurted out, "2, 5, 11, …"

$\mathcal{I}$n 1960 there was an NSF-sponsored conference at Dartmouth on the subject of function algebras—a very hot topic at the time. The atmosphere was quite relaxed. One day Matt Gaffney from the NSF showed up for a "site inspection". He could not find any mathematicians! In fact the only person he could find was one of the wives, and she was looking for her husband. It seems that all the mathematicians were out hiking in the countryside.

J. Wermer (1927–), K. de Leeuw (1930–1978), and S. Helgason (1927–) were out on a boat. Life was easy in those days, and there were no dire consequences. In today's competitive funding atmosphere, one wonders what the result might be.

*W*hen André Weil was a young man, he was a guest in Gösta Mittag-Leffler's villa in Djursholm, Sweden (now the home of the Mittag-Leffler Institute). He had a number of detailed conversations with the great man. They would always proceed thusly: The talk would begin in French, with some remarks on power series and Weierstrass's fondness for them. Then Mittag-Leffler would shift gears and begin speaking of his reminiscences of Weierstrass and Sonja Kowalevski; this was done in German. When Mittag-Leffler grew tired, he would begin to speak in Swedish. But then he would interrupt himself and say to Weil, "But I forgot you don't know Swedish. We will continue our work at another time." After a few weeks of this, Weil learned enough Swedish that he could follow all parts of his conversations with Mittag-Leffler.

*O*swald Teichmüller (1913–1943) was certainly one of the most brilliant and far-seeing mathematicians of the twentieth century. His ideas were way ahead of his time, and are still studied intensely today. But Teichmüller was always eccentric, sometimes crazy, and frequently exhibited instability and dangerous tendencies. He tried to burn down the chemistry building at his university. He published almost all his work in the Nazi math journal *Deutsche Mathematik*. He served in Hitler's army and died on the Russian front.

*R*obert Maynard Hutchins (1899–1977) is considered to have been one of the most successful and influential university presidents in American history. He was President of the University of Chicago from 1929 (beginning at the tender age of 30) until 1945. He then became Chancellor (a somewhat honorific position) for six years. He is remembered for having completely re-invented the U. of C. curriculum—including the creation of the Great Books program. Mathematicians remember Hutchins fondly because he allowed the Math Department to hire Marshall Stone. And he gave Stone free reign (and the necessary resources) to build up a first-rate mathematics department. It was Stone who hired S. S. Chern, A. Weil, I. Segal, and P. Halmos, among others. For a goodly time, the University of Chicago Mathematics Department was the best in the country, if not the world.

*R*obert Hutchins is remembered fondly by some of us for having said, "Whenever the urge to exercise strikes me, I lie down until it passes." He is also known to have said, "The only exercise I ever get is acting as pallbearer for my friends who exercise."

*P*aul Erdős was once on a plane, flying over Lake Superior. He was reading the Toronto newspaper. A particular passage perplexed him. He looked up from his reading and queried, "What is a Yuppie?" A nearby passenger began to explain the distinguishing traits of a "Young, urban professional." But Paul, in his threadbare shirt, with two days beard stubble and all his possessions in a carry-on bag, had no idea what the poor fellow was talking about.

On the same occasion, Erdős asked what a SONY Walkman was. A fellow passenger handed one over and told Paul to have a listen. He was enthralled, and commented that the sound was extraordinarily good. He wondered whether he should purchase one. In the end he decided it would be too much trouble to transport all the tapes. He would be content with his old and semi-functional transistor radio. "It's never really been a sacrifice for me. I never wanted material possessions." Paul concluded by saying, "Private property is a nuisance."

*I*n the late 1970s the Hollywood movie *It's My Turn* (originally called *The Eternal Triangle*) was released with great fanfare. It starred Jill Clayburgh as a mathematician at Columbia University and Michael Douglas as the baseball coach (and her lover). Benedict Gross (1950–)—then of the Princeton Math Department, today of Harvard—was the technical advisor to the film. Indeed, Clayburgh and her entourage spent several days in Princeton soaking up local color and learning "how a mathematician behaves." During that time, she went to lunch with the mathematicians, hung out at tea, and endeavored to learn all the right moves.

When the movie was actually being shot, they flew Gross out to Hollywood and put him up in the Beverly Wilshire Hotel. He was present on the set and gave of his wisdom to lend credence to a number of scenes. Actually, he tells of his first day arriving on the set. The security people did not want to let him in because it was clear that he didn't belong. One of the potentates had to intervene to get him access to the sanctum sanctorum.

There are two particularly memorable scenes in the movie. The opening shot has Clayburgh giving a lecture in an advanced graduate course. She is presenting the proof of the "fives lemma" from homological algebra. Dick Gross spent many hours teaching her this proof, and she finally got it right (although there is a graduate student in the class who keeps asking silly questions). The other notable vignette is when Clayburgh solves her research problem. She has been thinking hard about it, of course, throughout the film. But at one point she is putting on her makeup in the morning, looking in the mirror while she applies her eyeliner, and she suddenly realizes that if she *reflects* her construction then her problem is solved. Voilà! Life should be so simple.

"The Mac has a one-button mouse, the PC has a two-button mouse, and Linux has a three-button mouse. Is there a message in this medium? If I ever figure it out, I will let you know."

— Al Kelley

David Hilbert was great friends with physicist James Franck (1882–1964). One day Hilbert was out for a walk and he met Franck coming the other way. He said to Franck, "James, is your wife as mean as mine?" James Franck was rather taken aback by this query and didn't really know how to reply. He said, "Well, what has your wife done?" Hilbert said, "It was only this morning that I discovered quite by accident that my wife does not give me an egg for breakfast. Heaven knows how long this has been going on."

Sonja Kowalevski was of course one of the legendary women mathematicians. Her many accomplishments cause one to marvel that she died at the age of 41. In fact it is not well known that she was a talented writer. Her book *Recollections of Childhood* is but a slight indication of her considerable skills. One appreciator of Sonja's writing said, "The Russian and Scandinavian literary critics have been unanimous in declaring that Sophie Kovalevski [Sonja Kowalevski] was the equal of the best writers of Russian literature, in style as well as in subject matter." Her early death cut short her plans for various literary projects. In particular, she had planned to write "The Razhevski Sisters

Sonja Kowalevski

during the Commune," a reminiscence of her trip to Paris in 1871. Kowalevski found writing to be a cathartic escape from her intense periods of work on mathematics. She wrote in her memoirs that, "At twelve years old I was thoroughly convinced I was born a poet." She also recalls rather fondly her uncle buying math books for her, and speaking to her about the quadrature of the circle: "If the meaning of his words was unintelligible to me, they struck my imagination and inspired me with a kind of veneration for mathematics, as for a superior, mysterious science, opening to its initiates a new and marvelous world, inaccessible to the ordinary mortal."

*T*hat Kowalevski lived in Gösta Mittag-Leffler's home, and that they enjoyed an intense relationship (even though they were not married) is well known. An indication of the situation is given by Kowalevski herself: "Yesterday was a rough day for me, for big M [Mittag-Leffler] … left in the evening. If he had stayed here I do not know how I would have been able to work. He is so tall, so powerfully built, that he manages to take up a great deal of room, not only on a sofa, but also in my thoughts and I would never have been able in his presence to think of anything but him."

*I*n 1888 Kowalevski won the Bordin Prize for her paper "On the rotation of a solid body about a fixed point". This was a great triumph for her. She was in Paris, and was feted and invited everywhere. She was in her glory, and completely happy. Mittag-Leffler came to join her. However, the good times came abruptly to an end. With her demands and her tyrannical and jealous love, she expected a great deal of Mittag-Leffler. She thought that his admiration for her did not measure up to her love for him. Further, she was unwilling to give up her mathematical career and become simply the wife of this man whom she so loved and admired. Thus it was that they frequently separated; there was much bitterness and vituperation. Unable to live with him or without him, exhausted, torn by the incessant strife of their relationship, Kowalevski finally became ill and died in 1891.

*D*istinguished logician Saul Kripke (1940–) went to school in Omaha, Nebraska. When he was in eighth grade he solved a famous outstanding problem in mathematical logic. The result was published, and Kripke became an instant celebrity. One day a famous logician from Harvard phoned up Kripke to ask him some questions about his work. Saul's mother answered the phone. "Oh, Saul can't come to the phone right now. He is at Little League practice."

In later years, the high school principal at Saul's school used to like to console slow learners by pointing out that even Saul Kripke never learned to drive a car.

*W*e all like to eat, and mathematicians are no exception. Sometimes, however, we find ourselves out of our element.

Emil Artin, newly arrived in Bloomington, Indiana, once consulted the *Encyclopedia Brittanica* in order to learn how to carve a turkey. Carl Ludwig Siegel, for his dinner guest Harold Davenport (1907–1969), served a large trout for the first dinner course; then he served a large trout for the second course as well.

*A*ccording to Arnold Ross (1906–2002), Paul Erdős never had an academic job in the United States in part because administrators feared that he didn't know how to teach. Nonetheless, Ross arranged a position for him at Notre Dame. Erdős accepted with pleasure. At the first meeting of Ross and Erdős, Paul queried, "What will be my duties? You know I have not had a background of regular academic positions." Ross said, "Paul, you are a distinguished mathematician, so you must have a position that matches your distinction. You are going to be a full professor of mathematics, with all the duties and responsibilities thereto appertaining."

It was arranged that Erdős would teach an advanced course in set theory, and the graduate students who were to attend were handpicked. The students had the most exciting time of their lives: the course was challenging, but accessible. It was the first time they had seen real mathematics presented to them by a real research mathematician of world stature. As Ross put it, "Somehow, in the eyes of many near-sighted people, research and teaching are not related. We still have some people like that. If you knew Paul, you would understand that he would be an excellent teacher because his ideas of how to do mathematics were right."

$\mathcal{A}$s recounted in the first volume of *Mathematical Apocrypha*, Ludwig Bieberbach (1886–1982) was a Nazi sympathizer. He was also a racist and a bigot. On Easter Tuesday, 1934, Bieberbach gave a speech at the Technische Hochschule in Berlin entitled "Persönlichkeitsstruktur und Mathematisches Schaffen" [The Structure of Personality and Mathematical Creation] in which, to quote the newspaper *Deutsche Zukunft*, "he seems to show that the teaching of Blood and Race also applies here, and places the most abstract of sciences beneath the total state." In the same year Bieberbach published an article in the *Proceedings of the Berlin Academy* entitled "Stilarten Mathematischen Schaffens" [*Styles of Mathematical Creation*]. His point of departure seems to have been the SS-organized boycott of the lectures by the Jewish mathematician Edmund Landau the previous year. Courant wrote to Abraham Flexner (1866–1959) with this description of Bieberbach's ideas:

> The following event is a characteristic of the course of things. Professor Landau ... went to start his lectures last week. In front of his lecture hall were some seventy students, partly in S.S. uniforms, but inside not a soul. Every student who wanted to enter was prevented from doing so by the commander of the boycott [Werner Weber (1906– ?) who had once been Landau's Assistant]. Landau went to his office and received a call from a representative of the Nazi students, who told him that Aryan students want Aryan mathematics and not Jewish mathematics and requested him to refrain from giving lec-

tures …. The speaker for the students is a very young, scientifically
gifted man, but completely muddled and notoriously crazy. It seems
certain that in the background there are much more authoritative people who rather openly favor the destruction of Göttingen mathematics and science.

*I*n the nineteenth century many universities forbade women to enroll. At
one school, a talented woman persuaded the administration to allow her to
audit a math course. When the professor walked in the room on the first
day, he made his displeasure plain. He let it be known that the administration might be weak (to allow a woman to attend), but *he* would not change
his attitudes. He began his lectures, and made a point of proceeding at a fast
and furious pace. Week by week, this teaching style took its toll. By
midterm, only the woman and one struggling male student remained. The
professor did not let up one bit, and after a while longer the hapless male
student dropped out. The professor walked into the room to find only the
female student present to hear his lecture. Glancing around, the professor
announced, "Since there are no students here, I will not give a lecture." He
then walked out of the room.

*E*dgar Lorch tells of giving the "only course ever" at Columbia University
on mathematical logic. This was in the Summer of 1950. The heat was
unbearable, and all doors and windows were open. In the next room, Jean
Dieudonné was lecturing on group theory. Dieudonné's style was more
thundering than lecturing. Since Lorch's class had heard the entirety of
each of Dieudonné's lectures (as well as Lorch's), Lorch offered to let his
students take Dieudonné's final exam as well as his own.

*S*ome years ago a German professor of mathematics was visiting a large
midwestern university. Of course he was there primarily to conduct collaborative research with one of the faculty at that institution, but (in order to
subvent his living expenses) Herr Doktor Professor was assigned to teach
trigonometry and was given a syllabus and a rather weighty textbook on the
subject. At the end of the first week of classes, he marched into the under-

graduate office and declared, "OK, I have already covered this book. Now what should I do?"

A similar situation was encountered in the 1930s at the University of Pennsylvania when the Russian analyst Shohat was assigned to teach trig to freshmen. The chairman, J. R. Kline, was a quick thinker, and told the visiting professor, "Why don't you try to do it again?"

$\mathcal{D}$avid Blackwell (1919—), famous for his work in statistics and probability, spent a period at the Institute for Advanced Study at the beginning of his career. He remembers meeting John von Neumann on the first day, and von Neumann saying that Blackwell should come around and tell him about his thesis. Well, Blackwell was no fool. He figured that John von Neumann was just being polite, and really had no interest in his thesis. So he never went to tell him about it. But then, a few months later, their paths crossed again and von Neumann said, "When are you coming around to tell me about your thesis? Go in and make an appointment with my

David Blackwell

secretary." So, with some misgivings, Blackwell did. At the appointed time he went in to tell the great man of his modest contribution. John von Neumann listened for about ten minutes and asked a couple of questions, and then *he started telling Blackwell* about his thesis—what you have really done is this, and probably this is true, and you could have done it in a somewhat simpler way, and so forth. In Blackwell's words, "He listened to me talk about this rather obscure subject and in ten minutes he knew more about it than I did."

$\mathcal{A}$t MIT, Norbert Wiener met Ivan A. Getting (1912–2003), then a physics freshman and an organist, at a demonstration of a new electric organ. They became friends, and sometimes played tennis. On one occasion, Wiener failed to connect with 100 consecutive serves to him. Wiener then suggested that they might exchange racquets.

$\mathcal{E}$veryone knows that the first two Fields Medals were awarded to Lars Ahlfors (for work in complex function theory) and to Jesse Douglas (1897–1965, for work on the Plateau problem). Douglas could not be present at the International Congress of Mathematicians in Oslo to receive his prize, and Norbert Wiener stood in for him. Norbert really rose to the occasion, radiating personality, and he basked in the attention of dozens of Norwegian journalists and photographers. The Norwegians did not quite get the story straight, and they published

Jesse Douglas

pictures of Wiener with the caption that these depicted Jesse Douglas receiving his Fields Medal.

$\mathcal{J}$esse Douglas is remembered as a man who liked to tell amusing anecdotes and evoke witticisms. He once said that, as a rule, geometers are very nice people. But analysts are nasty.

After receiving the Fields Medal in 1936, Jesse Douglas was teaching at MIT. He suffered a nervous breakdown. There was a scramble to find someone to fill the slot, and Norbert Wiener's star pupil Norman Levinson was an obvious candidate. The story is told elsewhere of how G. H. Hardy intervened to make sure that Levinson got the job, even though he was a Jew.

Sadly, it took Douglas a great many years to recover from his illness. He spent the later years of his career as a teacher in New York City.

$\mathcal{R}$ecalling his early days as a mathematician, Armand Borel (1923—2003) says that, "During the summer, … I [went] to the first AMS Summer Institute, devoted to Lie algebras and Lie groups (6 weeks, about thirty participants, roughly two lectures a day, a leisurely pace unthinkable nowadays) and then to Mexico (where I lectured sometimes in front of an audience of one, but not less than one, as Siegel is rumored to have done once in Göttingen …).”

There is a story of Hilbert giving lectures to a class which was frequently attended only by one student. One day no students showed up, yet Hilbert could still be seen giving his lecture. A colleague questioned him about this afterwards, and he said, “Well, I’m certainly not going to cancel my lecture for these idiots!”

$\mathcal{T}$here was once a garden party at the Statistical Institute in Calcutta. Norbert Wiener was in attendance. Masani was standing near a table when someone (unknown to Masani) approached to pick up some refreshments. Wiener introduced himself and got the reply, “I am Abraham Matthai.” Wiener replied, “Matthai, that’s the name Matthew in Malayalam.” Dr. Matthai was a statistician. Thinking that Matthai had just met Wiener for the first time, Masani approached him afterward to tell him something of Wiener and his lectures and his exploits. Matthai laughed: it had been his third encounter with Wiener, and the third time he had learned about “Matthew.”

$\mathcal{I}$n the summer of 1995 I spent a very pleasant month visiting Australian National University as the Richardson Fellow. This is a delightful honorary position whose only formal duty is to give a controversial public lecture. I decided to lecture on fractals. They took me out to dinner beforehand. The distinguished chair professor Neal Trudinger (1942–) was to attend. But he fell victim to the copious Australian bureaucracy. What does this mean?

I spent most of my first week in Australia visiting various government and university offices and filling out forms. This just seems to be the nature of the place. But I figured that this was the price I had to pay for being a distinguished visitor. Turns out that the natives had to pay the piper also. One of the laws in Australia is that the cushion in your office chair must be pumped up once every two years. Trudinger’s was pumped up the day of my famous lecture. He went to sit down and fell off his chair. Because of his injuries, he was unable to attend the dinner.

*W*hen on the faculty at Penn State, I once fielded a phone call from a company called Center Engineering. The fellow on the phone told me he had a math question. It was this: "Suppose you have an angle. How do you find its decimal expansion?" I said, "*What*???" Even though I really wanted to be helpful, I had no idea what he wanted.[*] I said, "I'm not quite sure what you mean. Can you give me an example?" The immediate reply was, "No!" So much for communicating mathematics to the public.

*H*alsey Royden spoke very highly of Charles Loewner (1893–1968), one of the stalwarts of the Stanford Math Department. In particular, "He was a popular teacher of graduate students and an excellent dissertation supervisor. While at Stanford he probably directed more PhD students than the rest of the department's faculty combined. Lipman Bers, who wrote his dissertation for Loewner at Prague, once remarked that any mathematics department containing Loewner was fully qualified to give the PhD degree, even if he were the sole member! Carl [Charles] treated his students as colleagues, inspiring the best to superior work, while exhibiting much patience and help for the slower student. Because of the generosity of his help when needed, it has been said of his students 'the weaker the student, the stronger the dissertation.'"

Charles Loewner

*F*or the past thirty or forty years, young mathematicians have become accustomed to following a rather meteoric career path:

---

[*] I'm now a bit longer in the tooth, and I think that I now know what he needed. In classical texts, knowing an angle is knowing its tangent. He wanted a way to produce the arctangent of a given number. But the software package he was using—perhaps a version of `Fortran` or `BASIC`—did not have arctangent in its library of functions.

Instructor → Assistant Professor → Associate Professor → Professor. The most talented individuals traverse the entire gamut in well less than ten years. Such was not always the case. Halsey Royden tells what the career path of a "typical" mathematician was like a hundred or so years ago:

George Abram Miller (1863–1951):
  A.B., Muhlenberg College, 1887;
  PhD Cumberland University, 1892;
  Professor of Mathematics, Eureka College, 1888–1893;
  Instructor in Mathematics, University of Michigan, 1893–1895;
  Student, Universities of Leipzig and Paris, 1895–1897;
  Instructor, Cornell University, 1897–1901;
  Instructor in Mathematics, University of Chicago, summer 1898;
  Assistant Professor, Stanford, 1901–1902;
  Associate Professor, Stanford, 1902–1905;
  Professor, University of Illinois, 1906–1931.

Miller had a distinguished career, penning over 800 papers. He was a member of the American Academy of Arts and Sciences and the National Academy of Sciences. He served as President of the second organization.

Miller was a modest man. People were rather surprised when he bequeathed $1 million to the University of Illinois Mathematics Library. He said, "Everything I have I received from the university, and I want to repay my obligation." One consequence of this astonishing gift is that the University of Illinois has one of the best mathematics libraries in the country.

*S*olomon Lefschetz reminisced about his early career after obtaining his PhD from Clark University in 1911. "My first position was an assistantship at the University of Nebraska (Lincoln), soon transformed into a regular instructorship. This meant my first contact with a regular midwestern American institution and I enjoyed it to the full. I owed it mainly to the very pleasant and attractive head of the department, Dean Davis (1857–1918) of the College. The teaching load, while heavy, did not overwhelm me since it was confined to freshman and sophomore work.... Not too many weeks after my arrival, the Dean got me to speak before a group of teachers in Omaha on 'Solutions of algebraic equations of higher degree.' And then and there I learned an all-important lesson. For I spoke three quarters of an hour—three times my allotted time! When I found this out some weeks later from the Dean, my horror knew no bound. I decided 'never again,' to which I have most strictly adhered ever since."

𝒯he story of R H Bing's (1914–1986) name was included in the first volume of *Mathematical Apocrypha*. Now we include some additional information that comes from Bing's son-in-law Steve Hannah.

Bing's father was a school superintendent in a tiny east Texas town. His name was Rupert Henry Bing. And his mother was Lula May. The Bings had emigrated from Canterbury, England. When the future topologist was born, Rupert Henry announced to Lula May that their boy would be named "Rupert Henry." To which, according to Steve Hannah, Rupert's "tougher-than-a-nickel-steak" wife replied, "Over my dead body."

They were deadlocked over the boy's name. In the end, to terminate the internecine strife, they agreed on "R H". Mathematician R H Bing never complained about his name, or lack thereof. But he found it irritating that various bureaucracies and agencies could never understand the situation. For example, apropos of the visa story in *Mathematical Apocrypha*, Bing's *TIME* magazine subscription would always arrive addressed to "Ronly Honly Bing."

𝒞athleen Morawetz (1923–) is a distinguished mathematician at the Courant Institute of Mathematical Sciences in New York City and holder of the National Medal of Science. She told me that she had always wanted to serve on jury duty, but she could never pass the *voir dire*. Whenever one of the attorneys would question her, the issue of what she did for a living always came up. She would reply, "I'm a Distinguished Chair Professor of Mathematics at the Courant Institute," and that was it. They didn't want any eggheads on the jury.

But Cathleen was determined. So next time she was called for jury duty,

Cathleen Morawetz

and the question of her occupation came up, she said, "I'm a teacher." Good enough. Cathleen Morawetz got to serve.

$\mathcal{I}$t is well known among mathematicians—a story that we all tell to our calculus students—that Carl Friedrich Gauss, when he was ten years old—stunned his schoolteacher by performing the sum $1 + 2 + 3 + \cdots + 99 + 100$ —which the teacher had given the class in order to fill up the afternoon—in a minute or two. What Gauss did was to observe that the sum of an arithmetic series is the number of terms multiplied times the average of the first and last term. The story has, however, been transmogrified with time. It is thought that the actual sum that Gauss was asked to calculate was

$$81297 + 81495 + 81693 + \cdots + 100899.$$

This sum can of course be calculated by the same method.

$\mathcal{W}$itold Hurewicz (1904–1956) lived in Princeton in the late 1930s. He was famous for his mathematical prowess, but also for his unworldliness and his complete lack of physical coordination. Hurewicz frequently aggravated his landlady by letting his bathtub run over. He tried to play baseball and tennis with the graduate students who lived near his apartment, but he could never hit the ball or run as needed. It is speculated that this lack of physical prowess led to Hurewicz falling off a pyramid in Mexico on an excursion with Lefschetz (also discussed in *Mathematical Apocrypha*).

$\mathcal{T}$here is a math teacher in the University of California at Davis freshman program who has filed a claim asking to be excused from teaching trigonometry. Her claim is that it runs afoul of her learning disability.

$\mathcal{I}$ was once teaching a two-term advanced course in harmonic analysis. For the first lecture of the second term I had prepared a sweeping overview of the subject. I discussed singular integrals, elliptic partial differential equations, the Fourier transform, generalized systems of conjugate functions, pseudodifferential operators, and many far-reaching topics. The presentation reached a sublime level of euphoria, and the entire class seemed to be caught up in it. Of course I knew all the members of the class, both because they had been in attendance in the first term and because they were my friends whom I saw on a daily basis in the department.

But something new had been added. There was a lovely young woman—who almost looked like a movie star—who was in rapt attendance at this boffo performance I was delivering. She frequently gave off passionate sighs to punctuate the main points I was making. I kept wondering who this creature might be.

At the end of class, she waited for the others to leave the room and then she approached me, presumably to introduce herself. She took a deep breath and queried, "Is this remedial trigonometry?"

David Eisenbud (1947–), now the Director of the Mathematical Sciences Research Institute in Berkeley and President of the American Mathematical Society, was twice Chairman of his former department at Brandeis University. When I became Chairman of my own department, Eisenbud observed to me that it really wasn't such a terrible job. It only takes about half your time. "The trouble," said Eisenbud, "is that it's every other minute."

H. H. Wu (1941–) of U. C. Berkeley tells that on the wall of a New York City subway station was scribbled this message: "I have just found a truly marvelous proof of Fermat's Last Theorem. But the train is coming and I don't have time to write it down."

Wu also tells this story of his teacher Warren Ambrose (1914–1995). At MIT, the distance between the main math building and the cafeteria, from door to door, is about 100 feet. Even so, the young graduate students back in the 1960s would don their overcoats to go to lunch when the weather was inclement. After going to lunch a few times with Warren Ambrose, however, they felt shamed into *not* using their overcoats—just because Ambrose never used one. More important, he gave a mathematical justification that they could not refute: "It is only epsilon exposure."

Utter Solipsism

Peter Jones (1953–) is now a Professor of Mathematics at Yale—recently the Chairman of the Department. He was a graduate student at UCLA when I taught there, and it immediately became clear that he had a special talent.

In fact Peter came to Los Angeles not to study mathematics, but to be with his girlfriend, who was in another graduate program at UCLA. Peter worked at an animal hospital. But he decided to take Paul Koosis's graduate real analysis course. He came up with a clever new proof of the assertion that if $E \subseteq \mathbb{R}$ has positive Lebesgue measure then $E + E$ contains an interval. Thus Paul recognized that this student had a special talent, he was admitted to the graduate program with a T.A.-ship. His career took off like a rocket after that.

When Peter was set to take the qualifying exam in complex analysis, the writers put a problem on the exam that is equivalent to the Riemann hypothesis. They figured there was just a chance that Peter would pull off a real coup. Unfortunately, Peter did not solve the problem. None of the other students solved that problem either.

As is well established in story, fable, and movie, Alfred Dreyfus (1859–1935) was prosecuted for treason in the French military courts. Mathematician Henri Poincaré was called upon to testify on Dreyfus's behalf regarding questions of probability. In the course of things, Poincaré was asked why he was qualified to give testimony. He replied that he was the world's greatest expert. Asked afterward why he had said *that*, Poincaré said, "I was under oath. I could not lie."[*]

---

[*] It is said that Frank Lloyd Wright gave the same testimony—to the effect that he was the world's greatest architect—at a trial in Marin County, California.

In fact Dreyfus was a distant cousin of Jacques Hadamard's wife. As a result, Hadamard became actively involved in Dreyfus's defense. After he was acquitted, Dreyfus was awarded the Légion d'Honneur by the French government. Hadamard had considerable influence over that event. And Hadamard's abiding interest in left-wing politics—one which he maintained for the remainder of his long life—can be traced back to the events involving Dreyfus.

*R*ichard Bellman's (1920–1984) famous book *Dynamic Programming* is particularly noted for the collections of "Exercises and Research Problems" at the end of each chapter. Someone once asked Bellman how to tell the exercises from the research problems. He replied, "If you can solve it, it is an exercise; otherwise it's a research problem."

*F*ields Medalist Alain Connes (1947–) has spoken of the pain that accompanies mathematical discovery:

Alain Connes

> The process of verification can be very painful: one's terribly *afraid* of being wrong. Of the four phases, it involves the most anxiety, for one never knows if one's intuition is right—a bit as in dreams, where intuition very often proves mistaken. I remember once having taken a month to verify a result: I went over the proof down to the smallest detail, to the point of obsession—when I could have simply entrusted the task of checking the logic of the argument to an electronic calculator. But the moment illumination occurs, it engages the emotions in such a way that it's impossible to remain passive or indifferent. On those rare occasions when I've actually experienced it, I couldn't keep tears from coming to my eyes.

$\mathcal{G}$ian-Carlo Rota writes lovingly of being an undergraduate at Princeton in the early 1950s. The professors in the department, especially chairman Al Tucker, took an extremely paternal attitude in seeing that the good students went to the "right" graduate schools. Tucker sent Hyman Bass (1932–), Steve Chase (1932–), and Jack Egan to Chicago, Mike Artin (1934–) to Harvard, and kept John Milnor at Princeton. Rota decided for himself to go to the University of Chicago. He penned his letter of acceptance and was on his way with it to the mailbox when he ran into Chairman Tucker. Tucker asked Rota where he had decided to go for graduate study. When Rota told him, Tucker shouted, "You are not going to Chicago, you are going to Yale!" Rota tore up his letter and went to Yale to study with Jack Schwartz. He says it was the best decision of his life.

$\mathcal{T}$he poet Samuel Taylor Coleridge (1772–1834) is perhaps best remembered for his verse *The Rime of the Ancient Mariner*. The well-used phrase "an albatross hung about his neck" comes from that work.

On another occasion, Coleridge was moved to express the fact that it is possible, using only ruler and compass, to construct an equilateral triangle on any given base. The result is this:

> This is now—this was erst,
> Proposition the first—and Problem the first.
>
> I
> On a given finite Line
> Which must no way incline;
> To describe an equi-
> -lateral Tri-
> -A, N, G, L, E.
> Now let A. B.
> Be the given line
> Which must no way incline;
> The great Mathematician
> Makes this Requisition,
> That we describe an Equi-
> -lateral Tri-
> -angle on it:
> Aid us, Reason—aid us, Wit!

*H*ow does mathematical creativity take place? Where do good ideas come from? Of course there is no single answer, and as we all get older we wonder more and more why we don't have all the great ideas that we had thirty years ago.

There is general agreement, however, that one of the best ways to think about a problem is *not* to think about it. That is to say, you work hard and concentrate all your effort on the task for a protracted period of time; but then you let it go. Do something else: mow the lawn, go for a walk, take a shower, have a shave. The mind still percolates away, trying some things, discarding others. Often it happens that the solution to the problem will then just pop up unexpectedly.

Ron Graham, late of Bell Labs and now a Distinguished Professor at U. C. San Diego, gets some of his best ideas while juggling or performing gymnastics. Persi Diaconis (1945–) of Stanford lets his ideas gestate while he is creating and performing new magic tricks. Henri Poincaré used to go on trips to the seashore. Isaac Newton worked in his garden. Nobel Laureate Physicist Murray Gell-Mann (1929–) walks or cycles or cross-country skis. Fields Medalist Enrico Bombieri (1940–) paints. John E. Littlewood used to *always* take Sundays off. He absolutely forbade himself to think about mathematics on the day of rest. And then he always got a good idea Monday morning.

*A*lberto Calderón (1920–1998) was one of the great analysts of the twentieth century. He was a distinguished professor at the University of Chicago. Calderón was Argentine, and he spent as much time in the old country as he could. It was generally his habit to return to the University of Chicago (from Argentina) a week late into the semester—after classes had started. One year he phoned up the department secretary and said, "I'll be coming back on September 20." The secretary—who knew him well—said, "Oh, so this year you'll be on time for the semester, Professor Calderón." There was a pause on the line, and finally Calderón's voice re-emerged: "Then I'll come back on September 27."

*J*ohn Farey (1766–1826) was a geologist who lived during the Napoleonic era. It is unclear how he is remembered as a natural scientist, but he did make one decisive contribution to mathematics. In the paper "On a curious

property of vulgar fractions," *Philos. Mag. and Journal* 47(1816), 385—
386, he defined what we now call *Farey series*. These are series with the
property that they enumerate all fractions with denominator less than some
given integer k. For instance, the Farey series for $k = 6$ begins

$$\frac{1}{5}, \frac{1}{4}, \frac{1}{3}, \frac{2}{5}, \frac{1}{2}, \frac{3}{5}, \frac{2}{3}, \frac{3}{4}, \frac{4}{5}, \frac{1}{1}, \ldots$$

The ordering should be clear from this sample.

Farey observed that any fraction in the list (except for the first and last)
could be obtained by summing the numerators and denominators of the
fractions on either side. G. H. Hardy's commentary on the work can only be
characterized as acerbic:

> Just once in his life Mr. Farey rose above mediocrity, and made an
> original observation. He did not understand very well what he was
> doing, and he was too weak a mathematician to prove the quite sim-
> ple theorem he had discovered. It is evident also that he did not con-
> sider his discovery, which is stated in a letter of about half a page, at
> all important … He had obviously no idea that this casual letter was
> the one event of real importance in his life. We may be tempted to
> think that Farey was very lucky; but a man who has made an obser-
> vation that has escaped Fermat and Euler deserves any luck that
> comes his way.

Potter and claymaster Jasper Bond (1947–) told me that he went to high
school, in a suburb of Chicago, with celebrated logician Harvey Friedman
(1948–). Of course Friedman is a living legend. He appeared on the Ed
Sullivan television show in 1966 because he was the youngest person in history to earn a PhD—at the tender age of 18. He was then an Assistant
Professor at Stanford. Today Friedman is a Distinguished Professor at Ohio
State University, holder of the NSF's prestigious Waterman Prize.

Bond tells me that, when Friedman was in high school, his mental development had far surpassed his physical development. Harvey had few social
interactions, and he was extremely clumsy. But different people develop at
different rates, and Friedman has now surpassed most of us.

Analyst Ron DeVore (1941–) married young and had his children young.
So, in their early 40s, Ron and his wife found themselves footloose and

fancy free. They decided to build their dream house—on a lake, about 20 miles outside town. They designed the house themselves—every last detail—and Ron carefully supervised every step of the construction. He drove out to the site every day to check on progress, and he actually did some steps of the work himself—just to be sure that they were done correctly. Everything progressed slowly but surely until one day the electrician and the plumber were out there doing some work and ... the plumber pulled out a gun and shot the electrician dead. On DeVore's property! Construction was delayed for a year while all the nasty legal details were worked out.

*S*tanislaw Ulam was a remarkable presence and a considerable force in twentieth century mathematics. Although he could be arrogant and self-absorbed, his single-minded pursuit of mathematics is quite typical of those in the profession. Those reading his autobiography *Adventures of a Mathematician* [ULA] tend to find Stan quite charming in an Old World sort of way. But not H. E. Robbins (1915–2001). He thought that Ulam's book reflected rather badly on the profession and on Ulam. In his words:

> ... they are preoccupied with seeking recognition of their precise rightful place in the official pecking order. It is a pity that this aspect of the world of mathematics is so much emphasized in a book for the general reader; the more pity if indeed the emphasis is justified. The appearance of being thinking machines on the make, without discernible relation to parents, spouses, or children, and oblivious to the human concerns of our times, may be due in part to foreign systems of higher education that were devised to turn out idiot savants in the sciences as being more likely to be useful to the state. But if mathematical intelligence is strongly associated with emotional deprivation and social alienation, then even we earthy, super-honest, solid, and simple native Americans—the qualities that Ulam admires in us—are in for trouble.

G. H. Hardy is quoted (in *A Mathematician's Apology*) as saying that "good work is not done by 'humble' men. It is one of the first duties of a professor, for example, in any subject, to exaggerate a little both the importance of his subject and his own importance in it. A man who is always asking, 'Is what I do worthwhile?' and 'Am I the right person to do it?' will always be ineffective himself and a discouragement to others." One is reminded here of Norbert Wiener, who was famous for articulating exactly these self-doubts. Maurice Fréchet (1878–1973) says that, during his collaboration with Wiener, the great man was constantly saying, "Is my work worthwhile? Am I slipping?"

*O*ne of Hardy's perennial jokes was that he had an ongoing battle with God. God had it in for Hardy, and did his best to make Hardy's life miserable. One summer Hardy stayed in Engelberg, at a chalet owned by the Pólyas. It rained all the time that year, and as a result there were constant bridge games. The Pólyas, Hardy, and F. Gonseth (1890–1975) were constantly at the bridge table. After a while, Gonseth had to leave, and announced that his train was coming soon. Hardy said to Gonseth, "Please, when the train starts you open the window, you stick your head through the window, look up to the sky, and say in a loud voice: 'I am Hardy.'" The idea was that, once God thinks that Hardy has left, he will make good weather just to annoy Hardy.

G. H. Hardy

*S*oon after the University of Chicago was founded in 1892, E. H. Moore (1862–1932) was brought from Yale to establish the Department of Mathematics. Moore was a powerful personality and a great influence both on colleagues and students. He was meticulous in manners and dress. He would stop you in the hall, gently remove a pen from an outside pocket of your jacket, and give it to you—suggesting that you keep it in an inside pocket of your jacket. In those days, *everyone* wore a jacket. Moore was less cordial if you used your left hand as an eraser. He would become monumentally furious at any hint of intellectual dishonesty.

$\mathcal{F}$or all Gauss's many virtues, one cannot number kindness and supportiveness of young mathematicians among them. His consideration of Riemann is legendary, but he was less convivial with some others. For example, when János Bolyai's (1802–1860) father Wolfgang (1775–1856) presented Gauss with his son's remarkable construction of a non-Euclidean geometry, Gauss managed to abash both father and son with this reply:

> If I begin with the statement that I dare not praise such a work, you will of course be startled for a moment: but I cannot do otherwise; to praise it would amount to praising myself; for the entire content of the work, the path which your son has taken, the results to which he is led, coincide almost exactly with my own meditations which have occupied my mind for from thirty to thirty-five years. On this account I find myself surprised to the extreme.
>
> My intention was, in regard to my own work, of which very little up to the present has been published, not to allow it to become known during my lifetime. Most people have not the insight to understand our conclusions and I have encountered only a few who received with any particular interest what I communicated to them. In order to understand these things, one must first have a keen perception of what is needed, and upon this point the majority are quite confused. On the other hand, it was my plan to put all down on paper eventually, so that at least it would not finally perish with me.
>
> So I am greatly surprised to be spared this effort, and am overjoyed that it happens to be the son of my old friend who outstrips me in such a remarkable way.

János Bolyai was profoundly hurt by this condescending reply from Gauss, and it affected him for the rest of his life.

$\mathcal{J}$ack Schwartz was a student of Nelson Dunford, and Gian-Carlo Rota was a student of Jack Schwartz. Rota tells of how analysis at Yale in those days was completely dominated by *functional analysis* (Dunford and Schwartz being two of the principal exponents). At the same time, algebra was achieving an independent pinnacle of abstraction under the guidance of Nathan Jacobson and Øystein Øre. There was a standing bet among the graduate students of the time that whenever a doctoral dissertation in analysis was turned in, the writer would be challenged to use its results to give a new proof of the spectral theorem.

$\mathcal{A}$s indicated in the last story, abstraction was king at Yale in the 1950s. One distinguished mathematician pointedly remarked in 1955 that any existence theorem for partial differential equations which had been proved without using a topological fixed point theorem should be dismissed as applied mathematics. Another scion of the Yale corridors whispered to Rota in 1956, "Did you know that your algebra teacher Øystein Øre has published a paper in graph theory? Don't let this get around!"

Øystein Øre

$\mathcal{I}$n the mid-1970s, at the Joint Mathematics Meetings, there was a panel on what topics belong in an undergraduate abstract algebra course. The panelists were I. Herstein (1923–1988), MacLane and Jacobson. Mac Lane got up and outlined what sounded like a rather reasonable course, though, given his interests at the time, they were rather arrow-theoretic. Herstein did his turn with a description of what is more or less in his first abstract algebra text, a classic then as now. Finally Jacobson stood up and flashed as fast as he could the table of contents of his famous algebra text. It seemed to cover Noetherian rings rather quickly and go on from there. It looked like graduate school material to most of the audience. Finally, when the moderator asked if there were any questions or comments, Garrett Birkhoff got up and said that he would like to describe the course that he was teaching undergraduates at Harvard. But he first apologized by saying that from what he had just heard, what he was covering in his course would be known to every high school student in New Haven.

$\mathcal{J}$ohn von Neumann was once on a train and found that he was quite hungry. He asked the conductor to send the man with the sandwich tray to his seat. The busy and impatient conductor said, "I will if I see him." Johnny's reply was, "This train is linear, isn't it?"

$\mathcal{B}$rockway McMillan (1915–) tells of his interactions with Norbert Wiener and other denizens of MIT in the 1930s. He tells of the period 1938–1939 when Dirk Struik was giving lectures on the history of mathematics. This was a late-afternoon two-hour lecture once a week, not for credit. "Attendees came from all over. Wiener was one; he always sat in the front row next to the window. Regularly, behind him, sat a silent young man named Slutz. After an absence, Slutz reappeared with a handsome beard, a rarity among students in those days. Slutz also abandoned his habitual seat and sat in the center of the front row. Wiener arrived and sauntered across the front of the room toward his usual place. The new beard caught his eye. He sauntered back to the door and turned to survey the competition from a discreet distance. Then he walked briskly up to Slutz, thrust a welcoming hand forward and introduced himself: 'My name's Wiener.'"

McMillan relates that, somewhat later, Slutz had some notoriety in Princeton because he drove around in a Stutz sportscar. In the winter, he drained the radiator for the 5 or 6 days of his residence, then drove, with the cooling system dry (bereft of antifreeze), out to the Princeton traffic circle on U.S. 1 to fill the radiator there.

$\mathcal{J}$ulian Lowell Coolidge added to the luster of Harvard's mathematics teaching. After graduating *summa cum laude* from Harvard and then from Bailliol College in Oxford, Coolidge taught for three years at the Groton School. Then he returned to Harvard. Coolidge's particular passion was for geometry, and he taught a variety of courses and wrote supporting books. Coolidge's lively expository style is illustrated in his 1909 paper "The Gambler's Ruin." The paper concludes by reminding the reader of "the disagreeable effect on most of humanity of anything which refers, even in the slightest degree, to mathematical reasoning or calculation."

Coolidge was a vivid lecturer. He viewed research and scholarly publication as the *last* of the four major responsibilities of a university professor. According to Coolidge, these responsibilities were:

- To inject the elements of mathematical knowledge into a large number of frequently ill informed pupils, the numbers running up to 500 each year. Mathematical knowledge for these people has come to mean more and more the calculus.

- To provide a large body of instruction in the standard topics for a College degree in mathematics. In practice this is the one of the four that it is hardest to maintain.
- To prepare a number of really advanced students to take the doctor's degree, and become university teachers and productive scholars. The number of these men slowly increased [at Harvard] from one in two or three years, to three or four a year.
- To contribute fruitfully to mathematical science by individual research.

Coolidge was something of a celebrity. This was due in part to the fact that, during his time at Groton, he became great friends with our future president Franklin Roosevelt (1882–1945). At some point, years later, President Roosevelt penned a letter to Coolidge which ended, " … do you remember your first day's class at Groton? You stood up at the blackboard—announced to the class that a straight line is the shortest distance between two points—and then tried to draw one. All I can say is that I, too, have never been able to draw a straight line. I am sure you shared my joy when Einstein proved there ain't no such thing as a straight line!"

Bertrand Russell was once asked whether he believed in God. The immediate reply was, "Yes. Up to isomorphism."

Antoni Zygmund, whose name is now inextricably linked with the University of Chicago, was a very kind man and a very effective and successful mentor. He was not afraid to express strong opinions, however. On one occasion, when he was walking through several rooms in a museum filled with the paintings of a rather well-known modern painter, he mused, "Mathematics and art are quite different. We could not publish so many papers that used, repeatedly, the same idea and still command the respect of our colleagues."

Zygmund always endeavored to be kind in expressing judgments of mathematical colleagues. He is famous for the philosophy that he once expressed concerning the writing of letters of recommendation: "Concentrate only on the achievements, and ignore the mistakes. When judging a mathematician you should only integrate f^+ (i.e., the positive part of the function) and ignore the negative part. Perhaps this should apply more generally to all evaluations of your fellow men."

*J*n the early 1980s I wrote a paper about Sobolev spaces on the Heisenberg group. I submitted the paper to *Mathematische Annalen*. After a while, a very positive referee's report came back. The only criticism was that I spelled the spaces "Sobolov" rather than "Sobolev." I did so because my thesis advisor and mentor E. M. Stein used that spelling in his famous book *Singular Integrals and Differentiability Properties of Functions*. So I wrote back to the editor and told him I'd spell the name any way he wished, but I pointed out that a great authority, and also my teacher, had spelled it "Sobolov." No use. I had to spell it "Sobolev." I later learned that the referee of the paper was Stein himself. So I guess he got converted also.

W. Sierpiński (1882–1969) was always rather absent-minded. On one occasion, he and his wife were moving into a new residence. His wife knew that he was unreliable, so when they were down on the street with all their possessions she said, "Now you stand here and watch our ten trunks. I will go to get a taxi." She left him standing there, gazing off into space and humming absently. Some minutes later she returned and reported that she had called for a cab.

Sierpiński turned to his wife and, with a gleam in his eye, said, "I thought you said there were ten trunks. But I count only nine." The wife immediately became alarmed—panicked that some of their possessions might have been stolen from under her mathematician husband's nose. She cried, "No, there were supposed to be ten!" And she immediately began to count.

"No, no," said Sierpiński. "Just look: 0, 1, 2,…"

*R*ight after World War II, the Office of Naval Research was a principal source of funding for mathematics. Halsey Royden recalls some particulars: "One day the admiral in command of the Office of Naval Research (ONR) was rehearsing the staff at ONR for an impending inspection by the Chief of Naval Operations. He asked Jo (Joachim) Weyl (1915–1966), who had succeeded Mina Rees as the director of the mathematics branch, 'What do we tell the CNO when he asks why we spend all this money for research in mathematics?' Weyl responded with a typical metaphor: 'The tree of science has many branches, but the trunk is mathematics,' to which the Commandant said, 'No, no, much too high flown! We must have practical examples of the usefulness of mathematics.' Consequently, when the Commandant brought the CNO around to the mathematics branch, Weyl responded to the CNO's inquiries by talking about a number of research projects and pointing out their applications and usefulness to topics in physics and engineering. When Weyl had finished, the Commandant turned to the CNO and said, 'Perhaps I can explain it this way, Admiral: The tree of science has many branches, but the trunk is mathematics!' "

$\mathcal{L}$ipman Bers tells of how actions by the government—especially regarding funding—completely changed the character of American mathematics. To wit,

Mina Rees

After the war [i.e., World War II], government support of mathematics continued, primarily through the Mathematics Branch of the Office of Naval Research. (The part played by Mina Rees (1902–1997) cannot be overestimated.) In effect, for five years the ONR [Office of Naval Research] acted as the National Science Foundation which was founded in 1950. It established the system of summer grants, of support of graduate students, of support of conferences, and it developed the system of peer reviews. The effects, on mathematics, as well as on other sciences was dramatic and beneficial. Research was not anymore, as it used to be in all but a few elite institutions, the private pastime of professors paid for teaching elementary courses. Many universities were eager to hire mathematicians capable of doing research (and of obtaining a government grant). This changed the power structure in many universities giving the most qualified investigators the most influence on policy decisions. I am convinced that this by itself raised the intellectual level of many American universities.

We heard and we read much criticism of the grant system. We were told that professors' loyalties shifted from their institutions to their disciplines. In practice this meant, I believe, that good scientists became less dependent upon university administrators. We heard a lot about the evils of the "publish or perish" maxim, and much of what we heard is true. Yet this maxim sometimes replaced "serve on many committees or perish," "don't fail football players or perish" and even "go to the right church or perish."

At any rate, the grant system worked. Post-World War II America became the center of world mathematics.

J. L. Synge (1897–1995), father of Cathleen Morawetz of the Courant Institute, tells of meeting Oswald Veblen at an AMS meeting in Toronto in 1921. In an effort to make conversation with the great man, Synge asserted boldly that it was good to visualize things in Riemannian geometry. Veblen replied that visualization was completely useless.

Determined to prove Veblen wrong, Synge accosted Veblen later in the day and claimed that, in space-time, the Ricci tensor and the metric tensor defined a set of four principal directions. Veblen was impressed, and encouraged Synge to write up the result and get it published. Synge was pleased with this endorsement, and began to pursue the project. He soon discovered that the result was well known, but he ended up writing a more complicated and more general result. Veblen was true to his word and got the paper published in the *Proceedings of the National Academy of Sciences*. After that, Veblen and Synge remained good friends.

J. J. Sylvester once remarked that Cayley had been much more fortunate than himself: "that they both lived as bachelors in London, but that Cayley had married and settled down to a quiet and peaceful life at Cambridge; whereas he [Sylvester] had never married, and had been fighting the world all his days." Those in the know attest that this is a fair summary of their lives.

A mnam Neeman (1957–) was a Ph.D. student of David Mumford (1937–) at Harvard. Mumford gave Neeman a problem. Neeman solved it right away, but thought (mistakenly) that it was a practice problem. So he published the result in *Inventiones mathematicae* and did something else for his thesis.

David Hoffman (1944–), a geometer, was extraordinarily pleased when his son was admitted to Harvard for undergraduate school. He took the young man to help him get settled in his dormitory and to make sure he had everything he needed. A ritual excursion for all beginning freshmen is to go to Bob Slate's Stationery Store, just off Harvard Square, to stock up on folders and paper clips and other basic supplies. This is a tiny store, and only a couple of people can fit into the place at any one time. Utilizing the store for the same purpose while David and his son were there was another middle-aged man who was helping his daughter to get settled. Of course David had no idea who the other fellow was, but he seemed like a kindred spirit and they struck up a conversation. The gentleman was articulate and thoughtful and David found him to be a particularly charming conversationalist. When it was time to part, David offered his hand and said, "I'm David Hoffman. I'm a mathematician from Berkeley." His new friend said, "Glad to meet you. I'm Al Gore ..."

Norbert Wiener was a great mathematician, but he was plagued by self-doubt. He would question the janitors at MIT as to whether he was really the great mathematician that he was reputed to be. He would frequently go to the Chairman of the MIT Math Department and threaten to resign. Only after the Chairman (who in some instances was either Norman Levinson or William Ted Martin) offered his heartfelt reassurances of Wiener's enduring mathematical powers did the great man agree to stay on.

The International Congress of Mathematicians in 2002 was held in Beijing, China. One of the notable attendees was John Forbes Nash (1928–), recipient of the Nobel Prize[*] in 1994. Arrangements were made for Nash to give a special lecture (a technical treatise about game theory) one evening, and

---

[*] Technically, the "Economics Prize in Honor of Alfred Nobel," since Economics was not one of the fields specified by Nobel in his original trust.

it was a great hit. Many hundreds of people turned out to hear what Nash had to say. The next day my friend Christer Kiselman approached Nash and said how much he enjoyed the lecture. Christer commented that the attendance had been heartening. "Well," said Nash. "There was this movie you know (*A Beautiful Mind*, starring Russell Crowe) that seems to have attracted quite a lot of attention."

J was once discussing calculus reform with my friend Hugo Rossi (1934–) of the University of Utah. Hugo would tell me how they did things in Salt Lake City, and I would say that at Harvard they had a different take on the matter. Hugo would then tell me about an innovation that his Math Department had introduced, and I would reply that the Harvard Calculus book already contained the seed of that idea. Finally, in exasperation, Hugo blurted out, "Lightning can strike anywhere. But when it strikes at Harvard it does so with a Jovian bolt."

*J*n fact Hugo Rossi got mightily involved with the university administration of the University of Utah. He not only served as Chairman of the Mathematics Department, but at one point he became a Dean, and he later even rose to the eminent position of Vice President.

Hugo was Dean of Science during much of the low temperature fusion fiasco. Towards the end of it he was appointed Director of the Cold Fusion Institute. He became deeply involved in overseeing and, in the end, containing the situation.

You may recall that scientists Stanley Pons (1943–) and Martin Fleischmann (1927–) of the University of Utah announced, on March 23, 1989, that they had produced low-temperature fusion in their laboratory. Of course such a development could have mighty consequences for energy production, for defense, for the production of microprocessors, and the like. There was great excitement. Indeed, Pons and Fleischmann themselves were quite excited. Bypassing all the usual government agencies and funding sources, they went directly to Congress for funds to develop their ideas. In the end, this work was discredited—written off by some as just "dirty test tubes". The scientists were accused of fraud and incompetence. Pons and Fleischmann were humiliated by the scientific establishment, and they went into hiding.

But there was a brief moment during which the University of Utah was definitely in the limelight. Venture capitalists started showing up in Salt Lake City. Hugo recalls one particular day in which he was "taken for a ride" in a window-darkened limousine—by none other than future Clinton protege Ira C. Magaziner—and plied with a variety of proposals of a pecuniary nature. Of course each of the world's great religions has a vignette in which the prophet is tempted by mammon. Most of us who serve as university administrators are lucky enough not to be put in such a spot, however.

$\mathcal{C}$omplex analyst Halsey Royden of Stanford served the university in many different ways. At one point, in the early 1970s, he became a Dean. But he continued to study mathematics. And, during his tenure as an administrator, he managed to prove the remarkable theorem that the Teichmüller metric and the Kobayashi/Royden metric are always equal (whenever they are both defined). Royden liked to say at the time that this was the best theorem proved that year by a Dean.

Halsey Royden

$\mathcal{C}$laude Chevalley was already a quite distinguished mathematician when he spent a year at Harvard. He was assigned to teach calculus. After a time, one of the students came to the Department Chair to complain. He asserted that he did not understand a word of what Chevalley was saying, and neither did any of the other students. The student was sufficiently serious and articulate that the Chairman decided that the matter must be investigated. So he went to Chevalley's classroom at the appointed hour and discreetly stood in the hall in order to monitor what Chevalley was doing. To his surprise and chagrin, he found that Chevalley was teaching differential geometry—*not* calculus!!

Of course one could not go so far as to *correct* a man like Claude Chevalley. So the department set up a separate set of lectures where his students could go to learn calculus; Chevalley continued with his lectures on geometry.

One day I got a letter from my old graduate school friend Neal Koblitz (1948–). He recounted that his wife, Ann Hibner Koblitz, had been reading the text *Political Order in Changing Societies* by the celebrated political scientist Samuel P. Huntington (1927–) for her class in political theory. Huntington was Chair of the Government Department at Harvard, past president of the American Political Science Association, and an overall very distinguished man.

The book contained such formulas as

$$(1) \qquad \frac{\text{Social mobilization}}{\text{Economic development}} = \text{Social frustration}$$

$$(2) \qquad \frac{\text{Social frustration}}{\text{Mobility opportunities}} = \text{Political frustration}$$

$$(3) \qquad \frac{\text{Political participation}}{\text{Political institutionalization}} = \text{Political instability}$$

Huntington used this pseudo-mathematical analysis to reason that, in the early 1960s, South Africa was a "satisfied society."

Neal thought that such doubletalk was outrageous: What were the units? How could one take the measurements that would lead to equating such ratios? And was one allowed to use the laws of algebra to derive new formulas? Neal wanted to take action.

This whole matter resonated with me, because I had encountered similar reasoning (when I was a high school freshman) in an instructional movie hosted by Clifton Fadiman (1902–1999) about the novel *Great Expectations*. For those familiar with the story, Fadiman posited the equation

$$\frac{\text{Magwich}}{\text{Pip}} = \frac{\text{Miss Havisham}}{\text{Estela}}.$$

I was appointed by my English class to write a letter to Mr. Fadiman about the matter, and he and I ended up conducting a delightful correspondence.

I didn't really know what course to recommend to Neal Koblitz, but I had a feeling he had something in mind that was a bit more draconian than conducting salubrious correspondence with Clifton Fadiman. I suggested that he contact my friend Serge Lang, who is well known to wage battles and champion causes. The rest is a well-known phenomenon (and that history is recorded in charming detail in the book [LAN]). Lang took Huntington to task, proving categorically that in the 1950s and 1960s South Africa was the venue of egregious repression and social unrest. He excoriated Huntington—publicly—and he went so far as to block (twice!) Huntington's membership in the National Academy of Sciences.

Paul Erdős was passionate about his mathematics, and this passion sometimes led him to iconoclastic statements. For example, he criticized Andrew Wiles (1953–) for holing himself up for seven years in his attic to work on Fermat's last theorem (the FLT). Even Wiles's friend and collaborator Barry Mazur says that he had no idea what Wiles was doing during that time. Barry says that, one day, Wiles just came up to him, handed him a huge manuscript, and said, "Here's what I've been doing." It was his proof of the FLT.

Erdős said that the FLT "would have been solved a lot quicker if he [Wiles] had shared his work."

Once Paul Erdős was detained by the police for loitering. Paul did not have the experience nor the wherewithal to answer the police questions, nor to account for himself and his activities. He offered up one of his math papers—and they accepted it!!

*E*veryone knows that Paul Erdős was an itinerant scholar. He never owned nor rented a home, didn't have a driver's license, didn't have a credit card. He habitually would show up on the doorstep of a friend or a collaborator—anywhere in the world!—declare that "My brain is open."—and expect to be fed and housed and clothed. His motto was, "Another roof, another proof."

*O*nce Erdős had to have surgery on one of his eyes—for cataracts. He wanted, during the operation, to read a math book with the other eye. The physicians objected—for practical and scientific reasons both. As a compromise, the doctors arranged for a local mathematician to stay with Paul during the surgery and talk math.

*P*erhaps Erdős had traits in common with Leonhard Euler (1707–1783). Euler's life was completely devoted to mathematics. When he lost his right eye, he said, "Now I will have less distraction." Among Euler's many remarkable accomplishments was the ability to recite the *Æneid* from beginning to end, by heart. He even knew the first and last lines of each page of his edition of that classic text.

*E*rdős would let nothing stand in the way of his mathematical work. He was at breakfast, staying at a friend's house in New Jersey, when the name of a colleague in California came up. Erdős remembered a theorem that he wanted to tell the Californian about, and he headed towards the phone to place a call. His host interrupted, pointing out that it was 5:00AM on the West Coast. "Good," said Erdős. "That means he'll be home."

*E*rdős sometimes did mathematics like a chess grandmaster. He would have collaborators stationed around the room, or around the math department, and he would wander from person to person and room to room sharing thoughts. Sometimes he would be playing chess with one of the participants as well! Or he would read a book. He did not necessarily allow his companions to share in the same conceit. "No illegal thinking!" was his command when he sensed that someone's mind was wandering.

$\mathcal{L}$ouise Straus (1919–), who was married to Erdős's collaborator Ernst Straus and herself a mathematician, tells of life with Erdős:

> He'd just show up at our place, and we never knew how many days he was going to stay. I remember during the night hearing crashing sounds. The windows had no sash cords. If you opened the lock, they'd come crashing down. He ... could never figure out how to gently lower the windows ... (nor) how to manage the shower. He could never shut the faucets off. Water ran out on the floor. The linoleum buckled, and the door wouldn't shut again. He'd go outside to the pay phone and drop coins in it all night, calling mathematicians ... and asking friends to come over to our place ... He never asked us first if we wanted more guests.
>
> ...
>
> At parties, he always had trouble tying his shoes, ... I remember him sticking his foot out at the party, asking people to tie his shoe.

$\mathcal{O}$ne of the more bizarre Erdős quotations is "Louis the Fourteenth said, 'I am the state,' Trotsky could have said, 'I am society,' and I say, 'I am reality.'"

$\mathcal{E}$rdős is remembered for having said, "A mathematician is a machine that turns coffee into theorems." He was certainly a passionate advocate of, and drinker of, coffee. He once said, "In Hungary, many mathematicians drink strong coffee. At the mathematical institute they make particularly good coffee. When Posa [Lajos Posa (1949–), now a mathematician] was not quite fourteen, I offered him a little coffee, which he drank with an infinite amount of sugar. My mother was very angry that I gave the little boy strong coffee. I answered that Posa could have said, 'Madam, I do a mathematician's work and drink a mathematician's drink.' "

$\mathcal{W}$ilhelm Magnus (1907–1990) spent his younger days in Germany, in particular at the University of Frankfurt. He recalls that

The trouble with today's graduate students is that they don't smoke enough. We used to sit up all night smoking and talking mathematics.

Srinivasa Ramanujan (1887–1920) once said, "An equation for me has no meaning unless it expresses a thought of God."

$\mathcal{E}$rdős typically worked on mathematics—even when he was in a room full of collaborators—by putting his head in his hands and thinking as hard as he could. The others waited for Erdős to look up and offer an insightful comment about one of their problems. And often he would do this. But sometimes he would instead give an aphoristic statement having to do with death, such as, "Soon I will be cured of the incurable disease of life" or "This meeting, like life, will soon come to an end, but the meeting was much more pleasant." Then he would again bow his head and resume his meditations.

$\mathcal{M}$ost universities really like it if their faculty will attend the undergraduate commencement ceremonies. After all, the presence of a bunch of antiquated savants in traditional scholastic dress adds a certain decorum to the proceedings. But most faculty have neither the time nor the inclination to go. At Washington University in St. Louis, in the 1930s, the administration addressed this loggerhead as follows: each faculty member was handed his/her last paycheck *at the ceremony* as he/she crossed the stage. Amazingly, attendance at commencement was unanimous.

$\mathcal{I}$n 1996, Goro Shimura (1930–) of Princeton University was awarded the Steele Prize for Lifetime Achievement. In his written response to the award, Shimura took the occasion to wax philosophical:

> I always thought this prize was for an old person, certainly someone older than I, and so it was a surprise to me, if a pleasant one, to learn that I was chosen as a recipient. Though I am not so young, I am not so old either, and besides, I have been successful in making every newly appointed junior member of my department think that I was

also a fellow new appointee. This time I failed, and I should be grateful to the selection committee for discovering that I am a person at least old enough to have his lifetime work spoken of.

*S*himura is a great admirer of Carl Ludwig Siegel, but recalls him wistfully:

To clarify this point [that Carl Ludwig Siegel was disinclined to offer praise], we have to know what kind of man Siegel was. Of course, he established himself as one of the giants in the history of mathematics long ago. He was not known, however, for his good-naturedness. Around 1980 I sat next to Natasha Brunswick [Emil Artin's ex-wife] (1909–) at a dinner table, when she proclaimed, "Siegel is mean!" I don't remember how our conversation led to that statement, but many of those who knew him would agree with her opinion. Hel Braun (1914–1996), one of his few students, apparently disliked him.

He [Siegel] was indisputably original, and even original in his perverseness. Once at a party he played a piano piece and challenged the audience to tell who the composer was. Hearing no answer, he said it was a sonata by Mozart, Köchel number such and such, played backward. On the other hand, he had a certain sense of humor. When Weil asked him which work of his he thought best, he replied, "Oh, I think a few watercolors I made in Greece some years ago are pretty good."

*B*eing a mathematician can be frustrating. If you meet someone at a party, and he/she asks you what you do for a living, will you say, "I am a mathematician."? And what sort of answer can you then expect? A grimace and then, "Oh, I was always terrible at math." Or, "I was very good at everything up to algebra, but then all those letters confused me." As Lipman Bers said, "This is like a mathematician saying (about English), 'I was really good at A, B, C, but then those next twenty-three letters were really confusing.' "

Paul Halmos's take on the matter is as follows:

Do you know any mathematicians—and, if you do, do you know anything about what they do with their time? Most people don't. When I get into conversation with the man next to me in a plane, and he tells me that he is something respectable like a doctor, lawyer, merchant, or dean, I am tempted to say that I am in roofing and siding. If I tell him that I am a mathematician, his most likely reply will be that he himself could never balance his checkbook, and it must be fun to be a whiz at math. If my neighbor is an astronomer, a biologist, a chemist, or any other kind of natural or social scientist, I am, if anything, worse off—this man *thinks* he knows what a mathematician is, and he is probably wrong. He thinks that I spend my time (or should) converting different orders of magnitude, comparing binomial coefficients and powers of 2, or solving equations involving rates of reaction.

J was once at a cocktail party chatting animatedly with a quite charming woman. We were really hitting it off—up until the moment when she asked me what I did for a living. I said that I was a mathematician. She drew herself up haughtily and declared, "Then I'm sure we have nothing in common!" And she stalked away.

M olly Ringwald (1968–) is a talented young actress with flaming red hair and a pert and perky demeanor. She was once interviewed in the Sunday newspaper. Asked about her love life, she said, "Well, it's not really going anywhere these days. Of course you won't find me dating any mathematicians ...."

*J*t is fairly well known that Louis de Branges (1932–) proved the Bieberbach conjecture in 1984. The story pertaining thereto is quite interesting. First, the question: Call a holomorphic function f on the unit disc D *schlicht* if $f(0) = 0$, $f'(0) = 1$, and f is univalent. The conjecture is then that the jth coefficient a_j of the power series expansion of f about 0 satisfies $|a_j| \le j$.

The proof of de Branges comprised the entire second edition of his book on Hilbert spaces of entire functions. Given that Louis had announced solutions of other famous problems—such as the invariant subspace problem— that had been proved wrong, not too many of us were inclined to battle through this new proof. But providence intervened: de Branges spent a sabbatical in Leningrad, and the hard-working Russian mathematicians did indeed work through the proof and help de Branges distill it down to a mere 16 pages which now appear in *Acta Mathematica*.

Various people have worked on simplifying and extending de Brange's proof. One of the more remarkable efforts is due to Lenard Weinstein. His proof is only 4 pages; it begins with Loewner's differential equation—a very classical fact—and then uses *only calculus* to derive the conclusion. Unfortunately, however, this is some of the most condensed and difficult calculus that many of us have ever seen. No less a luminary than Walter Hayman (1926–) has produced a manuscript that works out all the details of the calculus. It is slightly longer than de Brange's proof in *Acta*.

Perhaps the most astonishing proof of the Bieberbach conjecture to date appears in a paper by Shalosh B. Ekhad and Doron Zeilberger (1950–) that was originally entitled "A wallet-sized, high-school-level proof of the Bieberbach conjecture." This paper—just over 2 pages—proceeds in two steps: First, one verifies the combinatorial identity

$$(1+w)\frac{d}{dt}\left\{\sum_{k=1}^{\infty}\left(\frac{4}{k}-kc_k(t)\overline{c_k(t)}\right)w^k\right\}$$

$$= (1-w)\sum_{k=1}^{\infty}\operatorname{Re}CT_z\left\{\frac{\frac{\partial f_t(z)}{\partial t}}{\frac{z\partial f_t(z)}{\partial z}}\cdot\left(2\left(1+\sum_{r=1}^{k}rc_r(t)z^r\right)-kc_k(t)z^k\right)\right.$$

$$\left.\cdot\left(2\left(1+\sum_{r=1}^{k}r\overline{c_r(t)}\,z^{-r}\right)-k\overline{c_k(t)}\,z^{-k}\right)\right\}w^k$$

using a computer; second, one checks that the coefficients $A_{k,n}(c)$ in the formal power series (Laurent in w) expansion

$$\left(1-z\left(2c+(1-c)(w+1/w)\right)+z^2\right)^{-1} = \sum_{n=0}^{\infty}\sum_{k=0}^{n}A_{k,n}(c)\left(w^k+w^{-k}\right)z^n$$

are nonnegative for $0 \leq c \leq 1$.

You know how stodgy journal editors can be. The official title under which the paper now appears is "A high-school algebra, 'formal calculus', proof of the Bieberbach conjecture [after L. Weinstein]." One should understand, of course, that "Shalosh B. Ekhad" is Zeilberger's nickname for his computer.

❧

$\mathcal{L}$ouis de Branges has been rather aggressive in making sure that he gets proper credit for proving the Bieberbach conjecture. There have been, as a result, some fairly large public fights. And some people feel that the

Russian mathematicians in Leningrad deserve a share of the glory. None other than Atle Selberg (1917–) has weighed in with an opinion on the matter:

> The thing is, it's very dangerous to have a fixed idea. A person with a fixed idea will always find some way of convincing himself in the end that he is right. Louis de Branges has committed a lot of mistakes in his life. Mathematically he is not the most reliable source in that sense. As I once said to someone—it's a somewhat malicious jest but occasionally I engage in that—after finally they had verified that he had made this result on the Bieberbach Conjecture, I said that Louis de Branges has made all kinds of mistakes, and this time he has made the mistake of being right!

$\mathcal{A}$ popular modern off-Broadway musical is entitled *Fermat's Last Tango*. It is quite extraordinary in that **(i)** it is about serious mathematics and **(ii)** it actually has lines that formulate serious mathematical thoughts.

Some examples are

> I knew, I swore,
> That elegant symmetry
> Of x squared plus y squared
> Is square of z
> Could not be repeated if n were three,
> Or more!

At one point, Fermat himself appears, singing

> Elliptical curves, modular forms,
> Shimura-Taniyama,
> It's all made up, it doesn't exist,
> Algebraic melodrama!

$\mathcal{L}$ouis de Branges is currently, and has for the last twenty years been, working on the Riemann hypothesis. Of course this is one of the Clay Institute Millennium Problems, so Louis stands to win \$1 million if he proves the theorem. He states that he would not accept the prize money. In detail,

> I believe there needs to be a mathematical institute in Bourcia [the French village from which de Brange's ancestors came], and the

needs of this institute [should] take priority over my needs, and therefore I would not accept the prize ... I also think that if I prove the Riemann Hypothesis I would be protected by society; that my university would take a different view of me, they would keep me longer, and that I would get benefits that would be more than the benefit of a million dollars. If I had a million dollars I would first of all be heavily taxed; I would be seen as somebody that would be working for the million dollars; there would be perhaps relatives of mine that would require help, and what benefit would that bring to my work, to have a million dollars?

One of the legendary moments in modern mathematics is the day, in 1972, when Hugh Montgomery (1944–) of the University of Michigan met Freeman Dyson of the Institute for Advanced Study. Of course Montgomery is a number theorist and Dyson is a mathematical physicist. What could they possibly have in common? Montgomery hadn't even really wanted to meet Dyson. But meet him he did, and Dyson politely asked Montgomery what he'd been thinking about lately. The reply was that he'd been thinking about the zeros of the Riemann zeta function. Montgomery

Freeman Dyson

went on to indicate a formula for their distribution that involved the expression

$$1 - [(\sin \pi u)/(\pi u)]^2.$$

Dyson paused a moment and then said, "Well, that's the density of the pair correlation of eigenvalues of random matrices in the Gaussian Unitary Ensemble."

Thus was established a connection between number theory and quantum physics. Ever since, the Riemann hypothesis guys have been studying random matrices. There is no evidence that the physicists are studying number theory.

$\mathcal{E}$ugene P. Wigner begins his famous essay "The unreasonable effectiveness of mathematics in the natural sciences" with this story:

There is a story about two friends, who were classmates in high school, talking about their jobs. One of them became a statistician and was working on population trends. He showed a reprint to his former classmate. The reprint started, as usual, with the Gaussian distribution and the statistician explained to his former classmate the meaning of the symbols for the actual population, for the average population, and so on. His classmate was a bit incredulous and was not quite sure whether the statistician was pulling his leg. "How can you know that?" was his query. "And what is this symbol here?" "Oh," said the statistician, "this is π." "What is that?" "The ratio of the circumference of the circle to its diameter." "Well, now you are pushing your joke too far," said the classmate, "surely the population has nothing to do with the circumference of the circle."

$\mathcal{J}$ohn G. Kemeny (1926–1992), in his essay "Rigor versus intuition in mathematics," tells an anecdote to illustrate the difference between rigor and intuition. It goes like this:

There was an advanced seminar in topology in which the lecturer devoted the entire hour to writing out a proof with complete rigor. After having filled all the blackboards, he had everyone in the room completely lost, including one of his own colleagues, who jumped up and said, "Look, I just don't understand this proof at all. I tried to follow you, but I got lost somewhere. I just didn't get it at all." The lecturer stopped for a moment, looked at him, and said, "Oh, didn't you see it? You see, it's just that the two spaces connect like this," intertwining his two arms in a picturesque fashion. And then his colleague exclaimed, "Oh, now I get the whole proof."

$\mathcal{K}$emeny goes on to tell another story about a very famous twentieth-century mathematician (whom he does not name, but we are left to suspect that it is Lefschetz) who was a co-founder of an important branch of mathematics. As Kemeny tells it:

He had published a certain paper in which he mentioned a theorem without proof, and a Russian mathematician wrote to him, asking whether it would be a possible to receive a proof. Our distinguished mathematician answered the Russian request. After about a month or so he had a reply. The Russian mathematician thanked him profusely; however, he had to point out that the proof sent was for a completely different theorem and that the proof was incorrect. As a matter of fact, this particular mathematician is credited with many incorrect proofs, and yet there isn't a single creative mathematician who would not list him as one of the greatest mathematicians of the century.

*M*y friends Raymond Ayoub (1923–) and Christine Ayoub (1923–) took a two-year leave, in the mid-1980s, from Penn State to spend time at the University of Saudi Arabia. There were many temptations to do so: Raymond is of Middle Eastern descent, it was exotic and exciting, and the money was good. Ray used to send me letters from time to time telling of his exploits. In one such, he recounted that he was helping the university to get its library organized. "They have lots of books, and unlimited funds," Ray reported. "But they don't seem to have the Dewey decimal system down pat. For example, they have two copies of your several complex variables book shelved between *Moby Dick* and *The Rise and Fall of the Roman Empire.*

*I*n the first volume of *Mathematical Apocrypha* we told tales of mathematician Peter Farkas (1948–). In fact his wife Donka Farkas (1952–) is an accomplished linguist, now on the faculty at the University of California at Santa Cruz. She tells of her struggles to get her PhD. She was living at home with her mother and grandmother, trying to be a good daughter and granddaughter while studying her socks off. One day she was sitting at the kitchen table working away when her grandmother said, "Donka, come have coffee with us." "No, no, Grandma. I must study." But grandma was not to be dissuaded. "Come on, Donka. You need a break. Have some coffee." "No, no, Grandma. My studies are important. I must work." Grandma would have none of it. "So you'll get your PhD fifteen minutes later. Have some coffee."

*G*arrett Birkhoff was a distinguished mathematician at Harvard University. He was preceded there by his father G. D. Birkhoff, who was arguably the first great American mathematician. Several years ago, there was a popular story among the graduate students at Harvard that one of them approached Garrett at the departmental Christmas party and asked, "Who's the greatest father-and-son team in mathematics?" Birkhoff modestly replied, "I don't know. Who?" The insolent reply was, "Gauss and his (barrelmaker) father."

$\mathcal{L}$ee Rubel was a distinguished mathematician at the University of Illinois who had vast and undifferentiated interests. On the sheet that listed faculty specialties, he listed over 50 areas of concentration. He had a very high-flown opinion of himself and his accomplishments.

In one particular year, the University was embroiled in a controversy over traffic on Green Street. In particular, there was a great conflict between foot traffic and automobile traffic. Rubel wrote a letter to the Urbana newspaper in which he said, "In my many travels around the world as a distinguished mathematician, I have observed a phenomenon that could be a solution to our problem. I don't know what they are called, but they are these small bumps that are put in the street ..."

$\mathcal{W}$hen Mark Green (1947–) and Jerry Folland were students at Princeton, they wrote a number of satirical songs about mathematics. One of my favorites is the *Ballad of John Milnor*, which begins in this way (to the tune of *The Ballad of John Henry*):

> When John Milnor was a little baby
> A sitting on his Daddy's knee
> He picked up a piece of paper, gave it half a twist
> Said π_1 of S^1 is $\mathbb{Z}$, lordy lord,
> π_1 of S^1 is $\mathbb{Z}$.

> Well John Milnor grew up quickly
> He was wise beyond his years
> He could calculate better than anybody else
> The differential equivalence classes of spheres, lordy lord.
> The differential equivalence classes of spheres.

> ...

$\mathcal{A}$nother favorite Folland/Green creation, to the tune of Simon and Garfunkel's *Richard Corey*, is the *Ballad of Alex Grothendieck*:

> They say that Alex Grothendieck's worth half of this whole town
> With integrable connections to spread his wealth around
> Born to categories, a child of Bourbaki
> He had everything a man could want
> Except simplicity
> But I, I work on my PhD
> And I curse the life I'm living
> And I curse my poverty
> And I wish that I could be
> Alex Grothendieck.
>
> The journals print his theorems almost everywhere he goes
> And he lectures all around the world about the things he knows
> And he's won a Fields Medal
> For the deepness of his thought
> I'm sure he must be happy, with all the things he's got
> But I, I work on my PhD
> And I curse the life I'm living
> And I curse my poverty
> And I wish that I could be
> Alex Grothendieck.
>
> His secretary Dieudonné, who does the writing up and such,
> Is grateful for his patronage and thanks him very much.
> So my mind was filled with wonder when in the news I read:
> Alex Grothendieck gave up math — and took up politics instead.
> But I, I work on my PhD
> And I curse the life I'm living
> And I curse my poverty
> And I wish that I could be
> Alex Grothendieck.

$\mathcal{F}$rançois Viète (1540–1603), philosophizing about mathematics in his book *Introduction to the Analytical Art*, said

> Finally the analytical art, having at last been put into the threefold
> form of zetetic, poristic, and exegetic appropriates itself by right to

the proud problem of problems, which is

TO LEAVE NO PROBLEM UNSOLVED.

*W*e have all had the experience of having papers or manuscripts rejected, but perhaps never with the charming dispatch of the following missive from an ancient Chinese economics journal:

> We have read your manuscript with boundless delight. If we were to publish your paper, it would be impossible for us to publish any work of lower standard. And as it is unthinkable that in the next thousand years we shall see its equal, we are, to our regret, compelled to return your divine composition, and to beg you a thousand times to overlook our short sight and timidity.

*H*aving trouble getting that latest mathematical effort published by your favorite journal? One wag offers this solution to publication woes:

> If you think that your paper is vacuous,
> Use the first-order functional calculus.
> It then becomes logic,
> And, as if by magic,
> The obvious is hailed as miraculous.

*F*ritz John had a decisive and pithy way of looking at mathematics. One of his aphorisms was, "Sometimes, if something is true, then it's not the right question."

*I*n a more puckish mood, Fritz John once said

> This equation has no practical significance. Nature has ways of getting around these things.

*T*he following article is alleged to have appeared in the *Chicago Tribune*, though nobody knows for sure:

News Item (June 23) — Mathematicians worldwide were excited and pleased today by the announcement that Princeton University Professor Andrew Wiles had finally proved Fermat's Last Theorem, a 365-year-old problem said to be the most famous in the field.

Yes, admittedly, there was rioting and vandalism last week during the celebration. A few bookstores had windows smashed and shelves stripped, and vacant lots glowed with burning piles of old dissertations. But overall we can feel relief that it was nothing—nothing—compared to the outbreak of exuberant thuggery that occurred in 1984 after Louis de Branges finally proved the Bieberbach Conjecture.

"Math hooligans are the worst," said a Chicago Police Department spokesman. "But the city learned from the Bieberbach riots. We were ready for them this time."

When word hit Wednesday that Fermat's Last Theorem had fallen, a massive show of force from law enforcement at universities all around the country headed off a repeat of the festive looting sprees that have become the traditional accompaniment to triumphant breakthroughs in higher mathematics.

. . .

Jerry Folland and Fields Medalist Charlie Fefferman penned the song *The Final Oral Exam*, to the tune of *There is Beauty in the Bellow of the Blast* from *The Mikado* by Gilbert and Sullivan. This was published in *The Mathematical Intelligencer* 14(1992), 31. It begins like this:

P: There is grandeur in the grading of a group,
There is beauty in a singularity;
 There's an elegance emergent
In an integral divergent
And the counting of the edges of a tree.

S: Yes, I love to practice counting
To enormous sums amounting
And especially on the edges of a tree.

P: Categories have a splendor that is grim,
And functors always soothe the troubled mind;
And to him that's mathematic
There is nothing too erratic
In a process that's stochastically defined.

S: Yes, despite its oscillation
I can find the expectation
Of a process that's stochastically defined.

P&S: If that is so, sing derry down derry,
It's evident, very, that $\genfrac{}{}{0pt}{}{\text{you are}}{\text{I am}}$ done.

Away $\genfrac{}{}{0pt}{}{\text{you'll}}{\text{I'll}}$ go with theorem and query

And corollary, till tenure's won.
Do you fancy you are erudite enough?
Information I'm requesting
On the subject I am testing:
Can you tell me more about the classic stuff?

…

$\mathcal{H}$ere are two examples of complex analysis in action, the first from Roger Penrose in "Physical space-time and non-realizable CR-structures" and the second from Charles Fefferman in "Parabolic invariant theory in complex analysis":

> The real and imaginary parts there are each, in effect, independent real analytic functions, so the freedom to be factored out by is that of four real functions of two real variables. The amount of intrinsic freedom in the structure of the boundary is therefore
>
> $$\frac{1 \text{ real function of 3 real variables}}{4 \text{ real functions of 2 real variables}}$$
>
> However, any finite number of functions of two variables must be regarded as "peanuts" in the context of free functions of three variables, i.e., it is completely swamped by the three-variables' worth of freedom, and makes no contribution to count.
>
> —Roger Penrose

> The key difficulty in this approach is that not all the variants of the Main Lemma are true. Therefore, at each stage we must be extremely careful to apply the correct reduction principle. Otherwise, we may ruin our chances by reducing something true to something false.
>
> The whole process may be viewed as a game: We want to prove the Main Lemma, while the devil wants to disprove it. We move by picking one of the four reduction principles, thus reducing the Main

Lemma to a finite list of simpler variants. The devil replies by picking one of the simpler variants (which he believes to be false) and challenging us to prove it. We reply by picking a reduction principle, so that again matters are reduced to a finite list of yet simpler variants. The devil replies by challenging us to prove one of the variants on that list, etc.

—Charles L. Fefferman

"All good Christians should avoid contact with mathematicians, for they and others of empty prophesies are in league with the devil and lead us into darkness." So spoke St. Augustine (354–430) about 1600 years ago. Some things never change. But it *should* be understood that, in Augustine's day, the word "mathematician" was used interchangeably with "astrologer".

The "Generals" are what the Princeton Math Department calls the qualifying exams. A student's Generals consists of him/her spending three hours locked in a room with three very distinguished mathematicians—quite an ordeal. There are no holds barred during the qual—the examiners can ask about anything. This scribe, for instance, was asked about a colloquium that had been given the week before by Charles Fefferman about his latest breakthrough.

The custom has been that those who pass the quals in any given Spring (and usually almost everyone passes) give a party for the entire department. It is usually held in the rather grandiose Professor's Lounge on the thirteenth floor of Fine Hall. Part of the tradition—in many years—is that a satirical play is written by the party-givers and is performed at this party. Many faculty and students are parodied, and a good time is had (we hope!) by all.

One of the really legendary skits was written by Jerry Folland and Mark Green. It was about the Mathematical Godfather (written at the time when Francis Ford Coppola's movie was a major hit). Don Spencer played the Godfather. In the opening scene (just as in Coppola's movie), the Spencer/Godfather figure sits in a chair and various supplicants come to him, kiss his ring, and ask for favors. One particular character humbly approaches the old man and says, "Godfather, I have been working on the Diophantine equation $x^n + y^n = z^n$. I want to prove that, for $n \geq 3$, there are no integer solutions." "That sounds like a good problem," says the Godfather. "But you see, Godfather, there is this guy across town named Fermat, and he is working on the problem too. I am very much afraid that he is going to beat me to it." The Godfather adopts a very serious look on his face and says, "You just tell me his address. I'll make sure it's his last theorem."

A later skit (in which this author was involved) starred two mathematicians who would bicycle from math department to math department, find out what problems were hot, and solve them for the grateful posers. These two talented individuals were known as the "bounding co-cyclists." The skit was a musical, and featured such hits as "Steal a Theorem from Milnor" and "The Lipschitz Condition Blues." John Milnor was in fact in the audience for the one and only performance of this masterpiece.

A solo effort by G. B. Folland, circa 1968, is *The Steenrod Square Canticle* from the *The H. Pétard Song Book*. It was first performed at Folland's "Generals Party" at Princeton in 1970 and was published in the *American Mathematical Monthly* 85(1978), p. 108. For those not in the know, it is sung to the tune of *Scarborough Fair* by Simon and Garfunkel.

> Are you going to Steenrod Square?
> (Nelson, Moore, Shimura, and Stein)
> Remember me to one who lives there,
> She once was a true love of mine.
>
> Tell her to find me a cardinal D
> (Nelson, Moore, Shimura, and Stein)
> That lies in between $\aleph_0$ and c,
> Then she'll be a true love of mine.
>
> Let her find me an integer $n > 2$
> And positive integers x, y, and z
> Such that $x^n + y^n = z^n$ is true,
> Then a true love of mine she will be.
>
> Tell her to bring me a manifold here,
> Closed, simply connected, of dimension 3,
> That is not a homeomorph of the sphere,
> Then a true love of mine she will be.
>
> Tell her to find me a proof that it's true
> That a non-real zero of $\zeta(z)$

Must have real part = one over two,
then a true love of mine she will be.

And when she's solved all these problems for me,
(Nelson, Moore, Shimura, and Stein)
We'll publish in the *Annals* for all men to see,
And she'll be a true love of mine.

*J*n his seminal work *A History of Western Philosophy*, Bertrand Russell offered this description of Pythagoras (569 B.C.–475 B.C.):

> He may be described, briefly, as a combination of Einstein and Mrs. Eddy. He founded a religion, of which the main tenets were the transmigration of souls and the sinfulness of eating beans. His religion was embodied in a religious order, which, here and there, acquired control of the State and established a rule of the saints. But the unregenerate hankered after beans, and sooner or later rebelled.
>
> Some of the rules of the Pythagorean order were:
> - To abstain from beans.
> - Not to pick up what has fallen.
> - Not to break bread.
> - Not to stir the fire with iron.
> - Not to eat the heart.
> - Not to walk on highways.
> - When the pot is taken off the fire, not to leave the mark of it in the ashes, but to stir them together.
> - Do not look in a mirror beside a light.
> - When you rise from the bedclothes, roll them together and smooth out the impress of the body.

*J*n the late 1980s, a group calling itself *Students Against Math 151* looted the offices of three University of Pennsylvania professors who were teaching the course. The raiders stole everything but the furniture, using laundry carts to remove at least 30 cardboard cartons of material. The miscreants, evidently students, left behind a note saying, "Students against Math 151."

The raids occurred a week after final grades in the course were posted. The course had 550 students and was taught by five professors. The offices of two of these professors were spared.

*T*om Lehrer (1928–) wrote *The Derivative Song* to the tune of *There'll be Some Changes Made*. It goes like this:

> You take a function of *x* and you call it *y*,
> Take any *x*-nought that you care to try,
> You make a little change and call it delta *x*,
> The corresponding change in *y* is what you find nex',
> And then you take the quotient and now carefully
> Send delta *x* to zero, and I think you'll see
> That what the limit gives us, if our work all checks,
> Is what we call *dy/dx*,
> It's just *dy/dx*.

*T*om Lehrer has been hailed far and wide as the quintessential satirical song writer. When he turned his pen to mathematics (after all, as the first volume of *Mathematical Apocrypha* tells us, he was a mathematics student), he was particularly poignant:

There's a Delta for Every Epsilon (Calypso)

> There's a delta for every epsilon,
> It's a fact that you can always count upon.
> There's a delta for every epsilon
> And now and again,
> There's also an *N*.
>
> But one condition I must give:
> The epsilon must be positive
> A lonely life all the others live,
> In no theorem
> A delta for them.
>
> How sad, now cruel, how tragic,
> How pitiful, and other adjec-
> Tives that I might mention.
> The matter merits our attention.
> If an epsilon is a hero,
> Just because it is greater than zero,
> It must be mighty discouraging'
> To lie to the left of the origin.

This rank discrimination is not for us,
We must fight for an enlightened calculus,
Where epsilons all, both minus and plus,
Have deltas
To call their own.

*A*nother Tom Lehrer effort is called *The Professor's Song*. This one is to the tune of *If You Give Me Your Attention* from *Princess Ida* by Gilbert and Sullivan. It starts out like this:

If you give me your attention, I will tell you what I am.
I'm a brilliant math'matician—also something of a ham.
I have tried for numerous degrees, in fact I've one of each;
Of course that makes me eminently qualified to teach.
I understand the subject matter thoroughly, it's true,
And I can't see why it isn't all as obvious to *you.*
Each lecture is a masterpiece, meticulously planned,
Yet everybody tells me that I'm hard to understand,
And I can't think why.

My diagrams are models of true art, you must agree,
And my handwriting is famous for its legibility.
Take a word like "minimum" (to choose a random word),*
For anyone to say he cannot read that, is absurd.
The anecdotes I tell get more amusing every year,
Though frankly, what they go to prove is sometimes less than clear,
And all my explanations are quite lucid, I am sure,
Yet everybody tells me that my lectures are obscure,
And I can't think why.

*T*oday computers are a part of everyone's life. They make many tasks easier and more efficient, but they can also chew up a lot of time—especially because so many features of the computer are so seductive. Nobel Laureate physicist Richard Feynman was one of the first to observe the latter feature—when he described what happened to the fellow who was supposed to

---

*In fact, when this song is performed live, the soloist writes "minimum" on the blackboard as an undifferentiated wiggly line. See the cognate story about Harald Bohr in *Mathematical Apocrypha*.

be managing the computer at Los Alamos during World War II (when Feynman was there with others, working on the atomic bomb):

> Well, Mr. Frankel, who started this program, began to suffer from the computer disease that anybody who works with computers now knows about. It's a very serious disease and it interferes completely with the work. The trouble with computers is that you *play* with them. They are so wonderful. You have these switches—if it's an even number you do this, if it's an odd number you do that—and pretty soon you can do more and more elaborate things if you are clever enough, on one machine.
>
> After a while the whole system broke down. Frankel wasn't paying any attention; he wasn't supervising anybody. The system was going very, very slowly—while he was sitting in a room figuring out how to make one tabulator automatically print arc-tangent X, and then it would start and it would print columns and then *bitsi, bitsi, bitsi,* and calculate the arc-tangent automatically by integrating as it went along and make a whole table in one operation.
>
> Absolutely useless. We *had* tables of arc-tangents. But if you've ever worked with computers, you understand the disease—the *delight* in being able to see how much you can do. But he got the disease for the first time, the poor fellow who invented the thing.

*R*obert Langlands (1936–) is one of the great mathematicians of our time. But even great minds can struggle with simple questions. Witness this interview from the *New York Times*, March 24, 1984:

Q: What does a theoretical mathematician do?

A: You mean what does he do or why does he do it?

Q: What is your aim, your goal?

A: Let me think a minute before I answer that question. Are you asking, what is the purpose of theoretical mathematics? What role does it play in the lives of mathematicians or what are the individual's motives?

Q: What does a theoretical mathematician do all day? What is the nature of his work? What is his pursuit, his activity?

A: All right, but of course there are many things one does. But you want to know what he does when he thinks?

Q: Yes, exactly.

A: Ah well, I've never been able to explain to anyone else, any non-mathematician, that to me at least the objects with which one deals are very real. But that's not precisely an answer to your question.

A friend of mine, who prefers to remain anonymous, struggled in the 1960s to figure out what a pseudodifferential operator is. The result was this poem:

> Oh without Atiyah-Singer do you think that they'd still linger?
> Though when Nirenberg's on fire there is much to awe-inspire
> And the people that came after have included Lars Hörmander,
> Even so, is it quite clear that we'd really have to fear
> That the things that have been shown would not otherwise be known?
> Oh the world loves structures dearly but it still does not seem that nearly
> Every theorem that's been prov-ed would not have to be remov-ed
> If we gave them up and went right on—alone.

*W*hat algebraists do and what analysts do are really quite different. The algebraist seeks *structures*, while the analyst seeks *estimates*. And the algebraist might wonder just what it is that the analyst wants to do with these estimates. We now provide two answers to that question, generously offered by two of the major analysts of the twentieth century:

> Estimating is the central activity in the theory of partial differential equations: it serves to justify both abstract existence proofs and numerical computations. Estimates, as often presented in a string of lemmas, may look singularly unattractive, lacking the elegance of giving the best constants, and merely concerned with orders of magnitude. They do, however express deep truths and lead to results not easily obtained by algebraic manipulations of the differential operators. The most complete estimates exist for differential operators … of the type called *elliptic*…
>
> —Fritz John (1910–1994)

> In the right hands, Schwarz's inequality and integration by parts are still among the most powerful tools of analysis.
>
> —Mark Kac (1914–1984)

The book *Partial Differential Equations for Scientists and Engineers* by Stanley J. Farlow (1937–) contains, in Appendix 2, a "PDE Crossword Puzzle." Some of the clues are:

Across

1. A PDE of the form $A(x,y) u_{xx} + B(x,y) u_{yy} = f(x,y)$ is called an _____ equation.
5. Types of equations where the principle of superposition holds.
12. When we solve Laplace's equation in polar coordinates by separation of variables, the ODE in r is named after this man.
29. The most basic idea behind the solution to linear problems is _____.

Down

2. An integral transform in which the kernel is a Bessel function.
4. A famous mathematician who died at the age of 27 and had nothing to do with PDEs.
9. Waves of the form $[\cos(n\pi t)\sin(n\pi x)]$ are called _____ waves.
20. Responsible for a very elegant solution of the one-dimensional wave equation in free space.

*E*very intensely focused discipline has its jargon, its "inside jokes," and its arcana. Mathematics is no exception. In 1994, in *The Mathematical Intelligencer*, this author published a "*Glossary of Common Math Terms*." Here are some excerpts:

Phrase	Meaning
This is trivial.	I forget the proof.
This is obvious.	You forget the proof.
This is a calculation.	Let's all forget the proof.
Send me your preprints.	Please go away.
Send me your reprints.	Please stay away.
Read my book.	I don't know.
That problem is intractable.	I can't do the problem so neither can you.
He's one of the great living mathematicians.	He's written five papers and I've read two of them.

What are some applications of your theorem?	What *is* your theorem?
I don't understand that step.	You goofed.
How do you reconcile your theorem with this example?	You're dead.
Your theorem contradicts my theorem.	I'm dead.
Where do you teach?	Do you have a job?
Your talk was very interesting.	I can't think of anything to say about your talk.
Have you had many students?	Do you have any social diseases?
I read one of your papers.	I wrapped fish with one of your papers.

Further Reading

[BEL] E. T. Bell, *Men of Mathematics*, Simon & Schuster, New York, 1965.

[BLO] S. Bloch, Review of *Étale Cohomology* by J. S. Milne, *Bulletin of the AMS*, new series, 4(1981), 235–239.

[COD] B. Beckman, *Codebreakers: Arne Beurling and the Swedish Crypto Program During World War II*, translated by Kjell-Ove Widman, American Mathematical Society, Providence, RI, 2002.

[BOA] R. P. Boas, Jr., *Lion Hunting and Other Mathematical Pursuits,. A Collection of Mathematics, Verse and Stories*, Gerald L. Alexanderson and Dale H. Mugler, eds., The Dolciani Mathematical Expositions 15, Mathematical Association of America, Washington, DC, 1995.

[CAH] Campbell and Higgins, *Mathematics: People, Problems, Results*, Wadsworth, Belmont, CA, 1984.

[DUR] P. Duren, *A Century of Mathematics in America*, with the assistance of Richard Askey and Uta Merzbach, American Mathematical Society, Providence, 1988–1989.

[ERD] P. Erdős, On the fundamental problem of mathematics, *Amer. Math. Monthly* 79(1972), 149–150.

[EVE] H. Eves, *Mathematical Circles 3 Volume Set*, The Mathematical Association of America, Washington, DC, 2004.

[EXO] G. Exoo, A Euclidean Ramsey problem, *Disc. Comput. Geom.* 29(2003), 223–227.

[FAD] C. Fadiman, *The Mathematical Magpie*, Simon and Schuster, New York, 1962.

[GOF] C. Goffman, And what is your Erdős number?, *Amer. Math. Monthly* 76(1969), 791.

[GRA] J. Gray, Did Poincaré say "Set theory is a disease."?, *Math. Intelligencer* 13(1991), 19–22.

[HAL] P. Halmos, *I Want to Be a Mathematician*, Springer-Verlag, New York, 1985.

[HAR] G. H. Hardy, *A Mathematician's Apology*, Cambridge University Press, London, 1967.

[JAC] A. Jackson, Comme Appelé du Néaut—As If Summoned from the Void: The Life of Alexandre Grothendieck, *Notices of the AMS* 51(2004), part I: 1038–1056l; part II: 1196–1212.

[KAC] M. Kac, *Enigmas of Chance*, Harper & Row, New York, 1985.

[KARL] Karl-Franzens-Universität Graz Institut für Mathematik, www.kfunigraz.ac.at/imawww/pages/humor/anekdoten_e.html.

[KTH] S. T. Kassouf and E. O. Thorp, *Beat the Market*, Random House, New York, 1967.

[KRA] S. G. Krantz, *How to Teach Mathematics*, 2nd edition, American Mathematical Society, Providence, 1999.

[LAN] S. Lang, *Challenges*, Springer-Verlag, New York, 1998.

[LIT] J. E. Littlewood, *Littlewood's Miscellany*, edited by Béla Bollobás, Cambridge University Press, Cambridge, 1986.

[MACH] D. MacHale, *The Book of Mathematical Jokes, Humour, Wit and Wisdom*, Boole Press, Dublin, 1993.

[MAL] Anecdotes about Mathematicians and Logicians, www.infiltec.com/j-logic.htm.

[MOR] R. E. Moritz, ed., *Memorabilia Mathematica*, Macmillan, New York, 1914.

[PAP] T. Pappas, *Mathematical Scandals*, Wide World Publishing, San Carlos, California, 1997.

[PAU] J. A. Paulos, *Mathematics and Humor*, University of Chicago Press, Chicago, 1980.

[POL] G. Pólya, *The Pólya Picture Album: Encounters of a Mathematician*, edited by G. L. Alexanderson, Birkhäuser, Boston, 1987.

[ROS] H. Rossi, ed., *Prospects in Mathematics*, American Mathematical Society, Providence, RI, 1999.

[ROT] G.-C. Rota, *Indiscrete Thoughts*, edited by Fabrizio Palombi, Birkhäuser, Boston, 1997.

[SAV] M. vos Savant, *The World's Most Famous Math Problem: The Proof of Fermat's Last Theorem and Other Mathematical Mysteries*, St. Martin's Press, New York, 1993.

[SCI] Science Jokes Collection,
www.xs4all.nl/~jcdverha/scijokes/joketalk.html.

[STAN] St. Andrews History of Mathematics Archive,
www-gap.dcs.st-and.ac.uk/~history/Mathematicians.

[THO] E. O. Thorp, *Beat the Dealer*, Random House, New York, 1962.

[ULA] Stanislaw Ulam, *Adventures of a Mathematician*, Scribner's, New York, 1976.

[VAN] B. L. Van der Waerden, *Algebra*, Ungar, New York, 1970.

[WEI] A. Weil, *The Apprenticeship of a Mathematician*, Birkhäuser, Boston, 1992.

[WHR] A. N. Whitehead and B. Russell, *Principia Mathematica*, Cambridge University Press, 1910.

[WIE1] N. Wiener, *Ex-Prodigy*, Simon & Schuster, New York, 1953.

[WIE2] ———, *I Am a Mathematician*, Doubleday, New York, 1956.

[WIL] L. Wilson, *The Academic Man: A Study in the Sociology of a Profession*, Oxford University Press, New York, 1942.

Index